for GCSE

Practice for book F1 part B

PATHFINDER EDITION

Contents

PUBLISHED BY THE PRESS SYNDICATE OF THE UNIVERSITY OF CAMBRIDGE
The Pitt Building, Trumpington Street, Cambridge, United Kingdom

CAMBRIDGE UNIVERSITY PRESS
The Edinburgh Building, Cambridge CB2 2RU, UK
40 West 20th Street, New York, NY 10011-4211, USA
10 Stamford Road, Oakleigh, VIC 3166, Australia
Ruiz de Alarcón 13, 28014 Madrid, Spain
Dock House, The Waterfront, Cape Town 8001, South Africa

http://www.cambridge.org

© The School Mathematics Project 2001
First published 2001

Printed in Italy by Rotolito Lombarda
Typeface Minion   *System* QuarkXPress®

A catalogue record for this book is available from the British Library

ISBN 0 521 01220 1 paperback

Typesetting and technical illustrations by The School Mathematics Project
Illustrations on pages 26, 27 and 28 by Chris Evans and on page 52 by David Parkins

16 Frequency

Sections A and B

1 This data shows the number of visitors to an exhibition each day in February.

24, 36, 41, 17, 25, 36, 22, 13, 49, 34, 16,

23, 19, 25, 32, 47, 37,15, 27, 29, 31, 28,

47, 39, 16, 21, 30, 19

(a) Copy and complete this stem and leaf table for the data.
(The first two pieces of data are already put in the table.)

```
0 |
1 |
2 | 4
3 | 6
4 |
5 |
6 |
```

Stem = 10 secs

(b) Copy out your table putting the 'leaves' in order.

(c) Find the range of the number of people.

(d) What was the median number of people ?

This table shows the data for March.

```
1 | 1 2 2 2 2 4 5 6 7 7 8 8
2 | 2 3 4 4 5 6 6 6 7 9 9
3 | 0 2 2 3 6 7 7
4 | 1 4
```

(e) Find the median and range of the number of visitors in March.

Section C

1 James is comparing two different types of tomato plants.
These tables show the number of tomatoes he was able to
pick from the different plants.

Many Maker   Tasty Toms

```
            | 1 |
        6 3 | 2 | 1
      7 4 2 | 3 | 4 5
    8 5 5 3 | 4 | 6 8 8
      4 2 4 | 5 | 3 7 9
          8 | 6 | 3
            | 7 |
```

Stem = 10 tomatoes

(a) Find the median and range of the number of
tomatoes for Many Maker.

(b) Find the median and range of the number of
tomatoes for Tasty Toms.

(c) Which of these statements is true ?

A The median for Many Maker is higher so they
produce more tomatoes on average.

B The highest number of tomatoes was picked from a
Tasty Toms plant so they produce more on average.

C The range for Tasty Toms plants was lower
so Tasty Toms produce more tomatoes on average.

Section D

1 The graph shows the ages of people who went
   to the *Visual Videos* on a Tuesday.

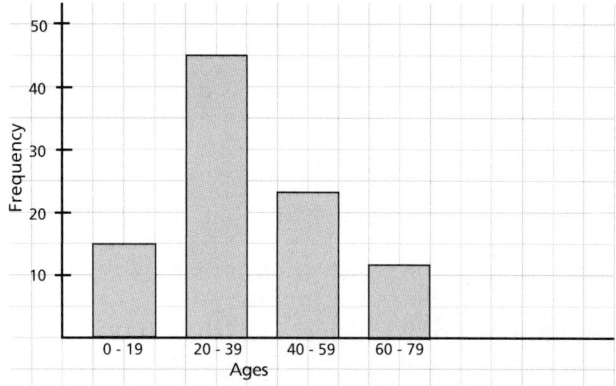

(a) How many people aged 40–59 went to *Visual Videos* on Tuesday?

(b) How many people aged 0–19 went to *Visual Videos* on Tuesday?

(c) What was the modal age-group for Tuesday?

(d) How many people in total visited *Visual Videos* on Tuesday?

2 This data shows the ages of people who visited
   Visual Videos on Saturday.

Age group	Frequency
0–19	25
20–39	44
40–59	51
60–79	10

(a) Copy and complete this bar chart to show the data.

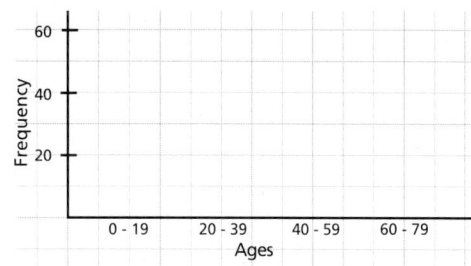

(b) (i)   How many people in total visited *Visual Videos* on Saturday?

   (ii) How many more people visited *Visual Videos* on Saturday than on Tuesday?

   (iii) Why do you think the numbers of visitors are so different on the two days?

(c) How many people over the age of 19 went to *Visual Videos* on Saturday?

(d) Give two differences between the people who went to *Visual Videos* on
   Tuesday and Saturday.

17 Number links

Section A

1 Which numbers in this list are multiples of 7?

    42, 24, 14, 17, 18, 700, 71, 35

2 Which numbers in this list are multiples of 3?

    23, 12, 9, 15, 31, 300, 28, 24

3 Write down six different multiples of 6 which are less than 64.

4 Write down six different multiples of 9 which are less than 100.

5 Write down all the multiples of 8 which lie between 30 and 50.

6 Which numbers in the loop are:

    (a) multiples of 7
    (b) multuples of 2
    (c) multiples of 11
    (d) multiples of 5
    (e) multiples of 4
    (f) multiples of **both** 5 and 4

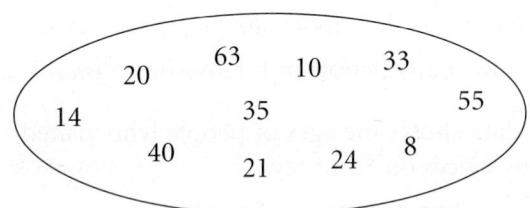

*7 6 is a multiple of 3 and is also a multiple of 2.
   We say that 6 is a **common multiple** of 3 and 2.

   Which of these are common multiples of 3 and 2?

   10, 12, 15, 18, 20, 24, 27

*8 15 is a multiple of 3 and is also a multiple of 5.
   Which of these are common multiples of 3 and 5?

    30, 24, 50, 45, 25, 33, 60

*9 List five common multiples of 5 and 4 which are less than 150.

*10 List five common multiples of 3 and 7 which are less then 150.

Section B

1 One number in this list is not a factor of 12. Which is it?

    2, 3, 4, 5, 6, 12

2 Two numbers in this list are not factors of 18. Which are they?

    2, 3, 4, 6, 8, 9

3 One factor of 24 is missing from this list. Which is it?

    1, 2, 3, 4, 6, 12, 24

4 Write down the factors of 7.

5 List all the factors of

  (a) 8   (b) 12   (c) 30

  (d) 16   (e) 13   (f) 22

6 Peter is trying to fit the numbers 2, 3, 5 and 9 into this 'factor grid'.

	is a factor of 12	is a factor of 45
is a factor of 18		
is a factor of 15		

Copy and complete the factor grid. You can only have one number in each square.

7 Show how you can fit the numbers 3, 4, 5 and 8 into this grid, one number into each square.

	is a factor of 24	is a factor of 20
is a factor of 40		
is a factor of 12		

*8 4 is a factor of 12 and also of 32.

We say that 4 is a **common factor** of 12 and 32.

Which of these are common factors of 12 and 32?

1, 2, 4, 5, 6, 8, 9, 10,

9 List the common factors of:

  (a) 12 and 18   (b) 10 and 15   (c) 8 and 20

Section C

1 Decide which of these statements are true and which are false.

(a) 3 is a factor of 12. (b) 9 is a multiple of 3.

(c) 15 is a factor of 5. (d) 5 is a factor of 25.

(e) 18 is a multiple of 6. (f) 24 is a multiple of 8.

(g) 16 is a factor of 4. (h) 12 is a multiple of 3.

2 There is only one way to fit all these nine numbers into the nine spaces in this grid.

Copy the grid and fill in the numbers.

12   24   3   9   36

20   32   18   15

	is a multiple of 2	is a factor of 24	is a multiple of 3
is a multiple of 4			
is a factor of 18			
is a factor of 60			

3 Which word, either 'factor' or 'multiple', should go in each statement?

(a) 4 is a .... of 24

(b) 7 is a .... of 28

(c) 15 is a .... of 3

(d) 9 is a .... of 27

(e) 9 is a .... of 3

*4 Decide which of these statements are true and which are false.

(a) 15 is a common multiple of 5 and 3.

(b) 4 is a common factor of 12 and 20.

(c) 3 is a common factor of 9 and 12.

(d) 28 is a common factor of 7 and 4.

*5 Write down two common factors of 24 and 16

*6 Write down two numbers less than 50 which are multiples of both 3 and 5.

Section D

1 List all the prime numbers between 20 and 30.

2 Four of the numbers in the loop are not prime.
Which are they?

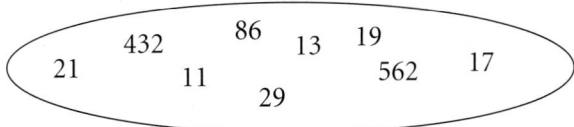

3 Which of the numbers below are prime?

12, 7, 48, 29, 5, 23

4 List all the prime numbers between 40 and 50.

5 Which of the numbers below are prime?

9, 12, 13, 17, 23, 25, 29

6 What is an easy way to tell that none of the numbers below are prime?

4, 56, 24, 128, 16, 34

*7 There is only one way to fit all nine numbers below
in the nine spaces in the grid.

Copy the grid and fill in the numbers.

9   4   7   2   3
18   1   8   6

	is an even number	is a factor of 36	is an odd number
is a factor of 24			
is a prime number			
is a factor of 18			

9

Section E

1 Which of the following are square numbers?

4, 6, 9, 18, 25, 81, 90, 100

2 Work these out.

(a) 5^2 (b) 4^2 (c) 8^2 (d) 11^2

3 Find each of these.

(a) $\sqrt{49}$ (b) $\sqrt{100}$ (c) $\sqrt{144}$ (d) $\sqrt{16}$

4 List the square numbers between 100 and 170.

5 This diagram shows a corner of a square made from 144 dots.

How many dots are along one edge?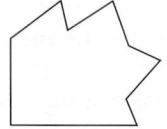

6 If a square is made from 169 dots, how many dots will there be along one edge?

Section F

1 Complete this list of cube number less than 150.

1, 8, 27, … , …… ,

2 Work these out.

(a) 7^3 (b) 11^3 (c) 8^3

3 Which of the following are cube numbers?

64, 25, 143, 125, 8, 9

4 (a) Find a cube number between 200 and 300.

(b) How many cube numbers are there between 300 and 400?

5 Work out the missing numbers.

(a) $\blacksquare^3 = 27$ (b) $\blacksquare^3 = 216$ (c) $\blacksquare^3 = 64$ (d) $\blacksquare^3 = 8$

Section G

Use the clues to find the numbers.

1
- A square number less than 100
- An odd number
- A multiple of 9

2
- A factor of 28
- A square number

3
- A cube number
- A multiple of 8
- Less than 10

4
- A prime number
- A factor of 28

5
- A prime number bigger than 8
- Less than 15
- A factor of 33

6
- A square number
- A factor of 27
- More than 1

7
- A prime number
- A factor of 20
- A factor of 15

8
- A factor of 18
- A factor of 24
- A multiple of 2
- A multiple of 3

9
- A factor of 24
- A square number
- More than 1

10
- A prime number
- More than 25
- Less than 30

11
- A multiple of 5
- A multiple of 6
- Less than 35

12
- A factor of 18
- A multiple of 9
- A multiple of 6

13
- A multiple of 7
- A multiple of 5
- Less than 40

14
- A cube number
- A factor of 24
- A factor of 16
- More than 1

15
- A square number
- A multiple of 6
- A multiple of 9

Section H

Find a way to number each number line.
(You may find that there is more than one way.)

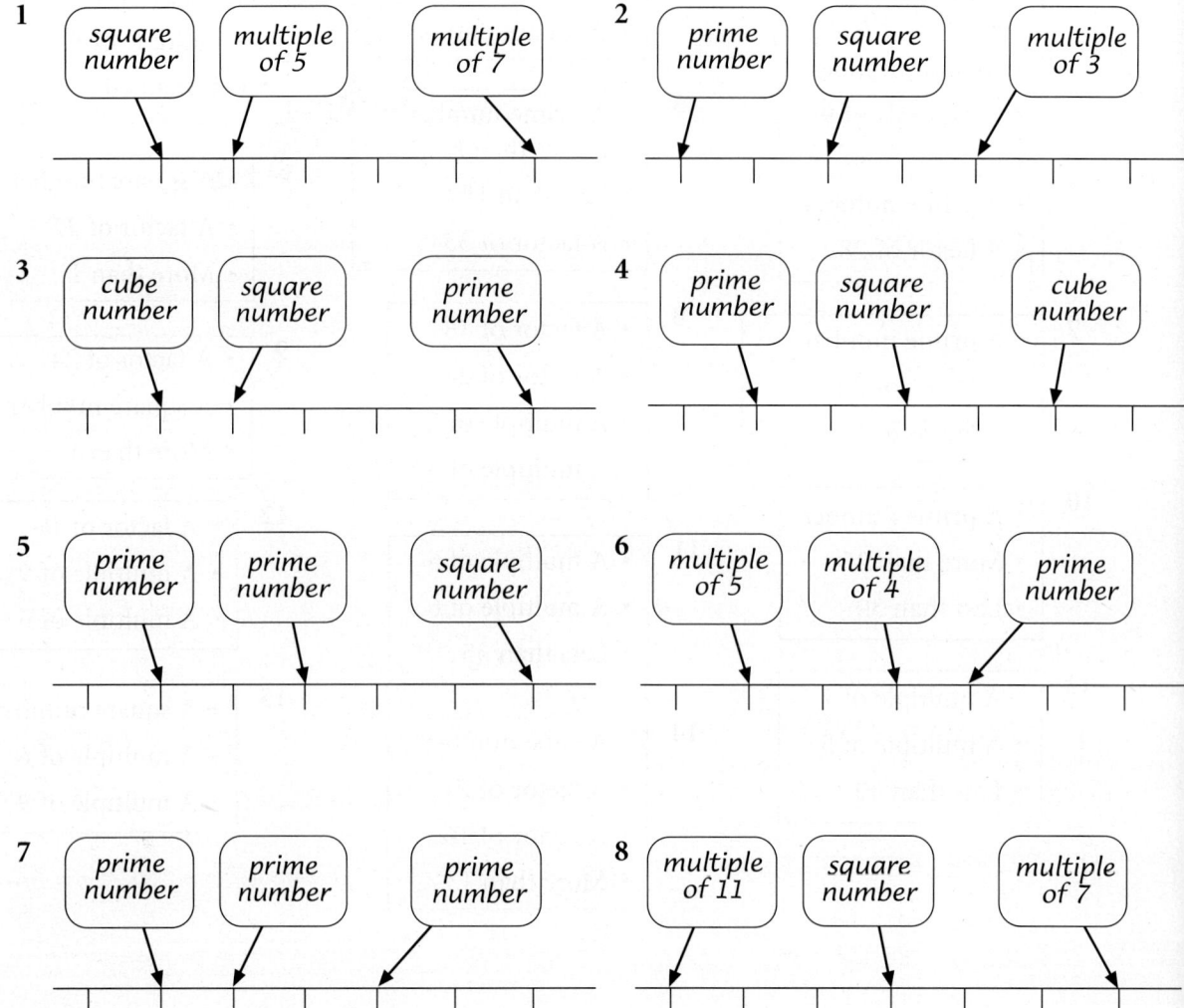

1
square number | multiple of 5 | multiple of 7

2
prime number | square number | multiple of 3

3
cube number | square number | prime number

4
prime number | square number | cube number

5
prime number | prime number | square number

6
multiple of 5 | multiple of 4 | prime number

7
prime number | prime number | prime number

8
multiple of 11 | square number | multiple of 7

Lines and angles

Section A

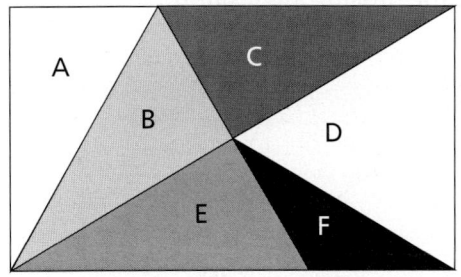

1 The diagram shows a rectangle made from triangles.

Describe each triangle as fully as possible.

2 These diagrams show parts of the rectangle above.

Calculate the missing angles.

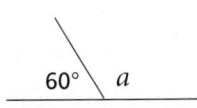 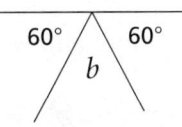 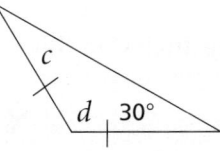 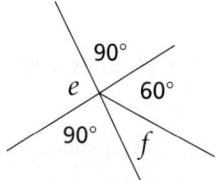

3 Calculate the missing angles

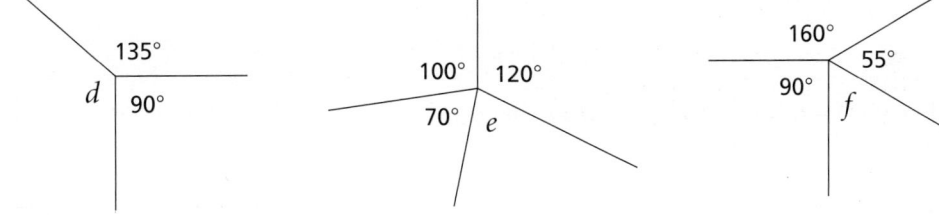

4 Calculate the missing angles in these triangles

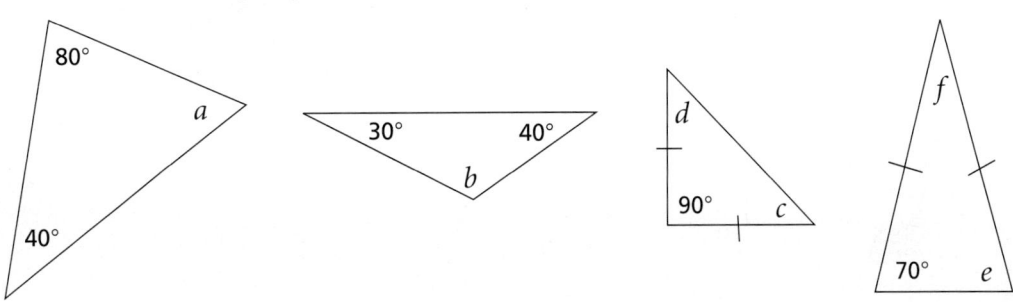

13

Section B

1 Calculate the missing angles in these diagrams.

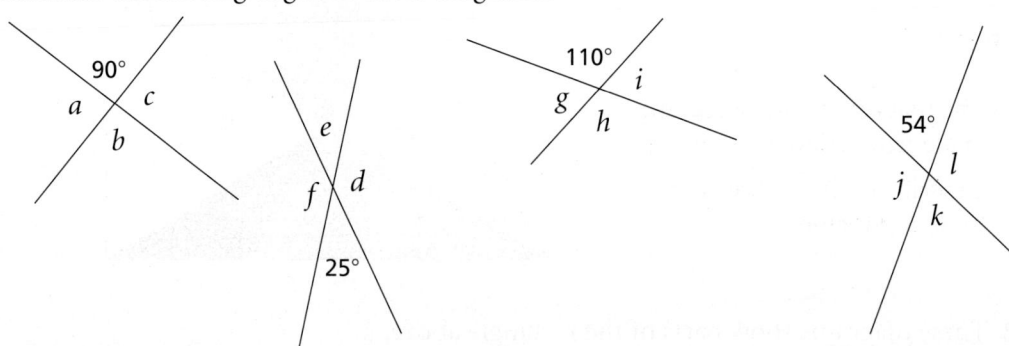

2 Calculate the missing angles on this cot, fishing stool and workbench.

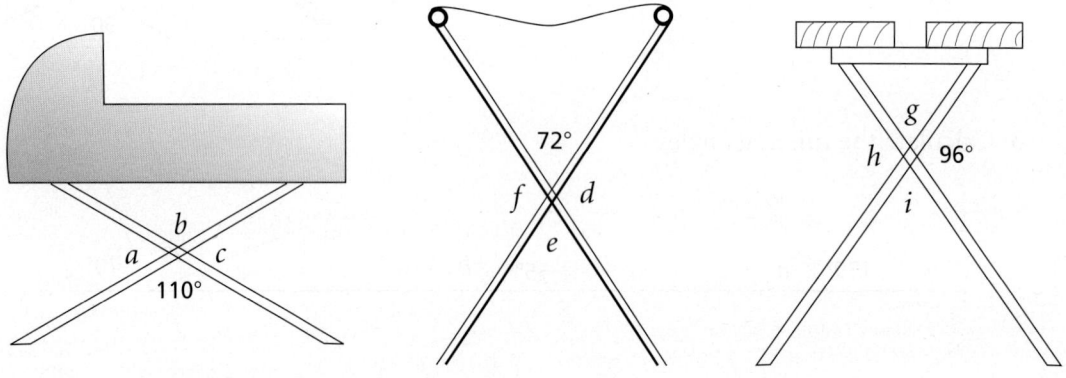

3 Calculate the missing angles in these diagrams

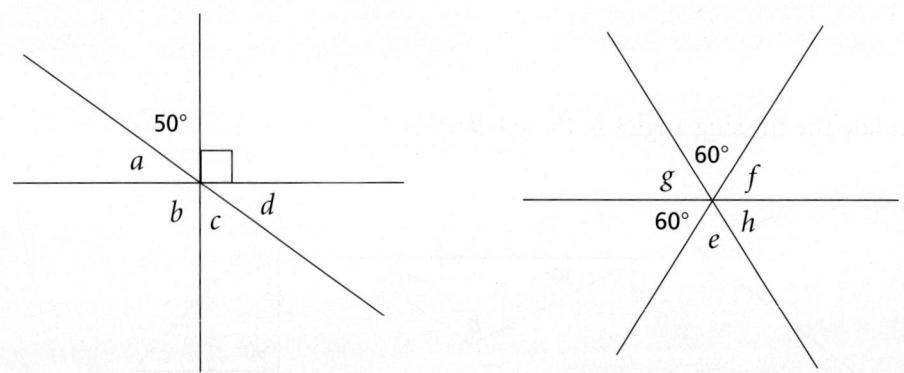

Section C

1 Find the missing angles in these diagrams

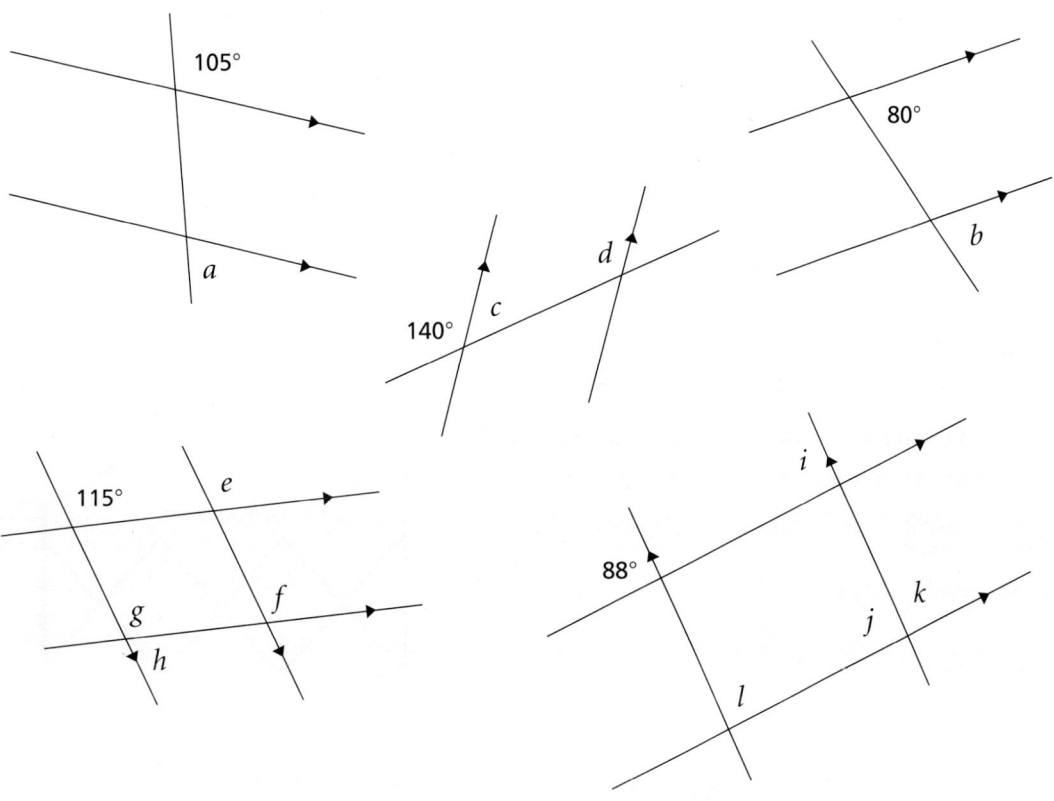

2 The picture shows a tree pruner.

  (i) Calculate angles *a* and *b*.

  (ii) Calculate angle '*a*' when the handle 'H' is pulled back so that angle '*b*' is 120°

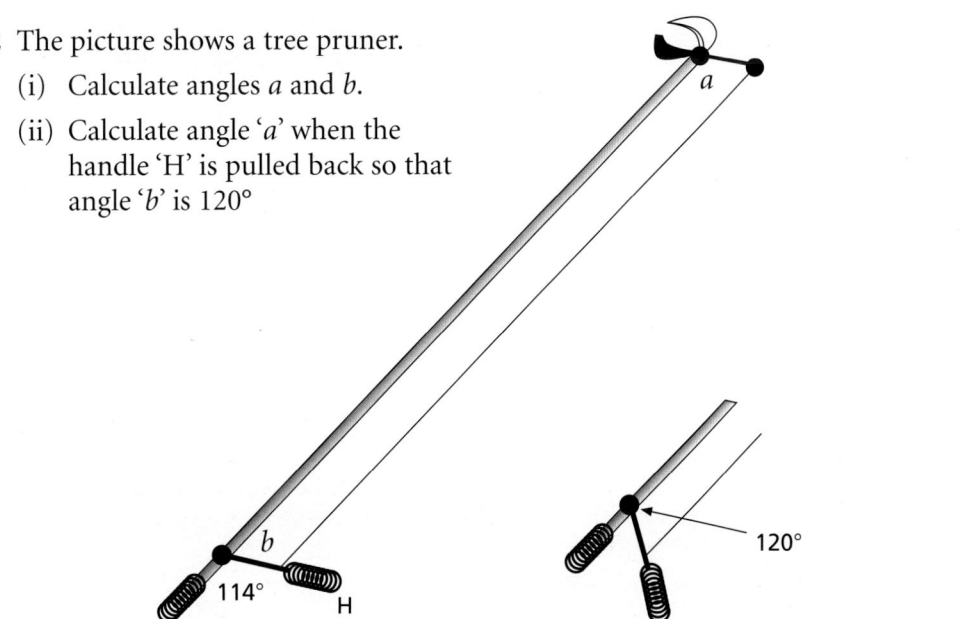

Section D

1 Find the missing angle
  in these diagrams.

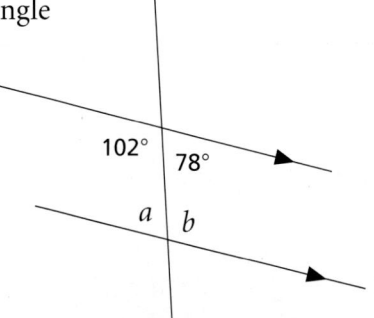

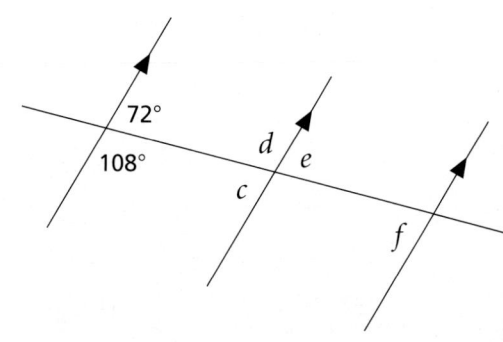

2 The picture shows a shaving mirror.
  Copy and complete these statements
  (a) Angles *a* and _____ are alternate angles
  (b) Angles *b* and _____ are alternate angles

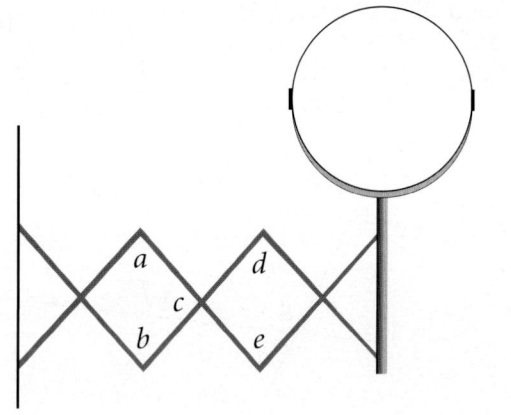

3 The pictures show a shed door and a gate.
  Find the unknown angles.

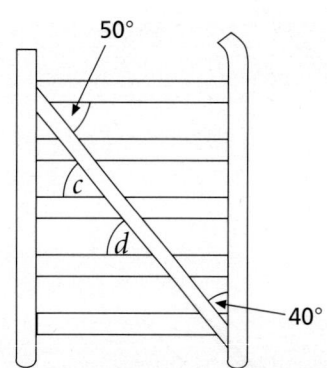

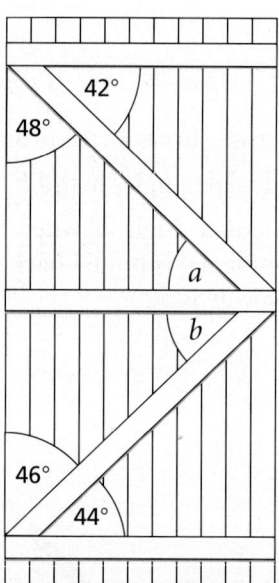

16

Section E

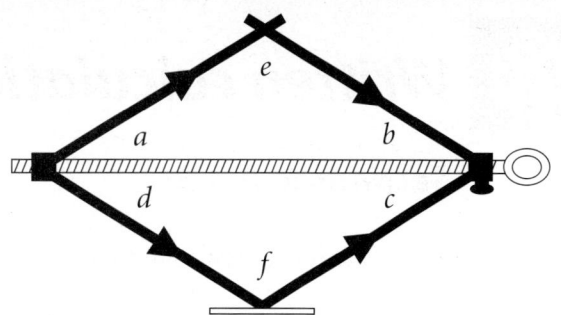

1 The sides of this car jack are parallel.

   (a) Which angle is alternate to angle *a*?

   (b) If angle *a* is 40°, and angle *b* = 40° find the unknown angles.

2

The diagram shows part of a lattice window.

   (a) Name three pairs of corresponding angles.

   (b) If angle *a* is 105°, find the unknown angles.

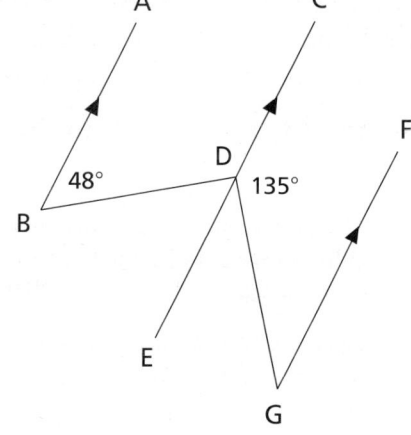

3 The lines AB, CE and FG are parallel.

  Find angles

  (a) BDE   (b) BDC   (c) GDE

  (d) BDG   (e) DGF

  giving reasons for your answers.

4 (a) What type of quadrilateral is BCDE?

  (b) Find with reasons the angles

     (i) CBE   (ii) BED

     (iii) ABE   (iv) BEA

     (v) BAE

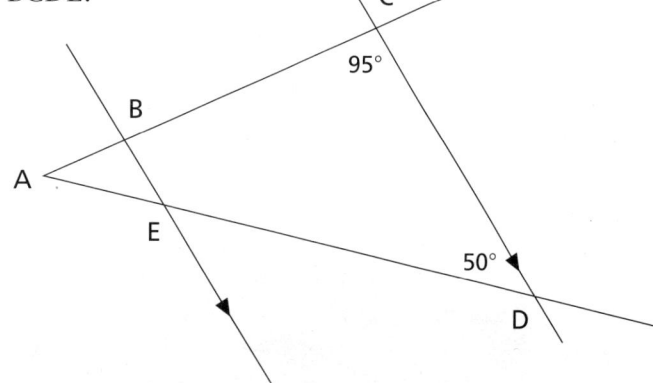

20 *Written calculation 1*

Sections A and B

1 (a) Make the largest total that
you can using the digits 3, 7, 6, 2.

(b) Make the largest *difference*
that you can using the digits 3, 7, 6, 2

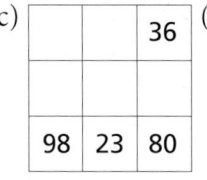

2 In a magic square the numbers in each **row**, **column** and **diagonal** add to the same total.
Copy and complete these magic squares.

(a)

11		21
	15	
		19

(b)

48		
28		44
32		

(c)

		36
98	23	80

(d)

3.6		
1.6	2.4	
2		

(e)

4.7		
2.1	4.8	7.5

3 Work out

(a) 36 + 48

(b) 156 + 385

(c) 509 + 294

(d) 83 – 48

(e) 126 – 72

(f) 431 – 275

4 (a) Here are four of the guesses of how many sweets there are in the jar.
The actual number of sweets is 834.
Which guess is closest?

916   783   870   796

(b) Here are four of the guesses for the weight of the cake.
Its actual weight is 3.67 kg.
Which guess is closest?

4 kg   3.25 kg   3.9 kg   3.4 kg

5 Work out in your head

(a) 3 + 2.6

(b) 4.8 – 2.3

(c) 8.1 + 3

(d) 4.6 – 2

(e) 4.6 + 5.4

(f) 5 – 0.4

(g) 1.5 + 1.25

(h) 2 – 1.2

6 Work out   (a) 4.5 litres + 1.25 litres   (b) 3 kg – 1.45 kg   (c) 3.7 – 1.24 m

7 Which basket of shopping is heavier and by how much?

A

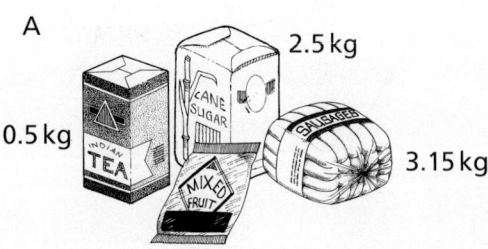

2.5 kg
0.5 kg
3.15 kg
1.76 kg

B

2.65 kg
0.35 kg
1.08 kg
3.25 kg

Section C

1 The perimeter of a shape is the distance around the outside.
  Find the perimeters of these shapes.

(a)
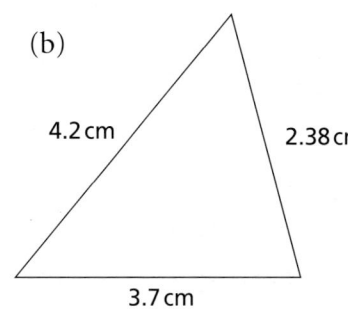
133 m

87 m

49 m

115 m

(b)
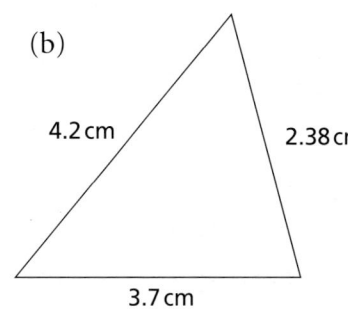
4.2 cm

2.38 cm

3.7 cm

(c)
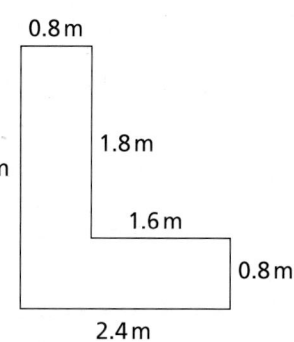
0.8 m

1.8 m

2.6 m

1.6 m

0.8 m

2.4 m

2 A triangular field has perimeter 560 m.
  One side has length 213 m and another has length 148 m. How long is the third side?

3 (a) Angles on a straight line add up to 180°.
      Find the missing angle.

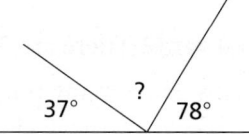
37°   ?   78°

  (b) Angles at a point make 360°.
      Find the missing angle.

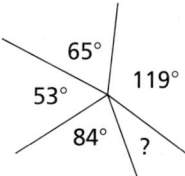
65°
53°   119°
84°   ?

4 (a) The sum of the angles of a triangle is 180°.
      Find the missing angle.

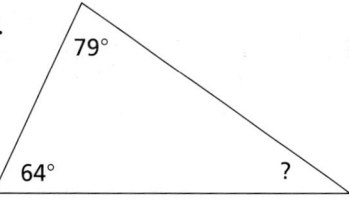
79°
64°   ?

  (b) The sum of the angles of
      a quadrilateral is 360°.
      Find the missing angle.

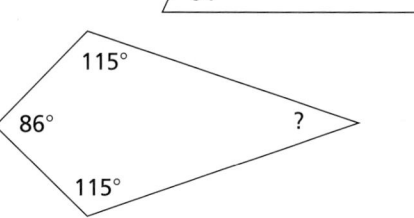
115°
86°   ?
115°

5 Here are the lengths of some leaves.

   3.6 cm   2.7 cm   3.9 cm   4.25 cm   2.85 cm

The range of these lengths is the difference between the longest and shortest lengths.
Find the range of the lengths.

19

Sections D and E

1 Work out

   (a) 38 × 5   (b) 142 × 4   (c) 215 × 8   (d) 467 × 9

2 Work out the answers to these then use the code to find the words.

0	1	2	3	4	5	6	7	8	9
P	A	T	C	H	W	O	R	K	S

   (a) 6 × 613   (b) 314 × 7   (c) 1852 × 5

   (d) 3155 × 3   (e) 4 × 8543   (f) 12553 × 3

3 A carton holds 9 packets of cereal.
How many packets of cereal will there be in 15 cartons?

4 Imperial units.

   (a) A mile is 1760 yards. There are 3 feet in a yard. How many feet are there in a mile?

   (b) A stone is 14 lbs. How many pounds are there in 8 stone?

   (c) There are 8 pints in a gallon. How many pints is 24 gallons?

5 Work out

   (a) 0.6 × 2   (b) 1.5 × 4   (c) 2.1 × 3   (d) 2.5 × 4

6 Choose numbers from the loop to
make these multiplications correct.

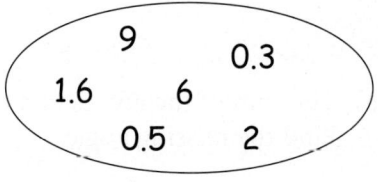

   (a) 0.9 × ☐ = 1.8   (b) 8 × ☐ = 2.4

   (c) 0.5 × ☐ = 4.5   (d) ☐ × 2 = 3.2

   (e) ☐ × 0.4 = 2.4   (f) 4 × ☐ = 2

7 How much will it cost to buy 6 cards at £1.40 each?

8 Some books are 2.3 cm thick.
How high will a pile of 9 of these books be?

9 Work out

   (a) 1.45 × 5   (b) 2.14 × 3   (c) 4.28 × 4   (d) 6.29 × 6

10 One kilogram is 2.2 lbs.
How many pounds weight is a 7 kg bag of potatoes?

11 A litre is 1.75 pints.
How many pints does a 5 litre container hold?

12 One inch is 2.54 cm.
How many centimetres is 8 inches?

Section F

1 Work out

(a) $85 \div 5$ (b) $76 \div 4$ (c) $189 \div 3$ (d) $836 \div 2$ (e) $924 \div 6$

2 Glass tumblers are sold in packs of 6.
How many packs can be made from 210 tumblers?

3 Nine classes need to share 234 books equally.
How many books will each class get?

4 Work out

(a) $4359 \div 3$ (b) $2528 \div 8$ (c) $1015 \div 7$ (d) $3807 \div 9$

5 The 1650 pupils at a school are divided equally into 3 houses for sporting activities.
How many pupils will there be in each house?

6 Find three matching pairs of division that give the same answer.
Which is the odd one out?

A $\boxed{136 \div 4}$ B $\boxed{875 \div 7}$ C $\boxed{4392 \div 6}$ D $\boxed{975 \div 3}$

E $\boxed{2196 \div 3}$ F $\boxed{1125 \div 9}$ G $\boxed{204 \div 6}$

7 Choco bars are packed in fives.

(a) How many packs of 5 can be made up from 84 bars?

(b) How many bars will be left over?

8 Match these in pairs where the remainder is the same. Which is the odd one out?

A $\boxed{239 \div 4}$ B $\boxed{326 \div 7}$ C $\boxed{493 \div 8}$ D $\boxed{324 \div 5}$

E $\boxed{590 \div 8}$ F $\boxed{471 \div 6}$ G $\boxed{761 \div 9}$

9 Oranges are sold in bags of 5.

(a) How many bags can be filled from 467 oranges?

(b) How many oranges will be left over?

10 Each table in a school dining room seats 8 people.

(a) How many tables will be needed for 675 people?

(b) How many empty seats will there be?

11 A car transporter can carry 7 cars.

How many transporter loads will be needed to move 1580 cars?

Section G

1 Work out

   (a) $3.6 \div 4$ (b) $7.2 \div 4$ (c) $8.6 \div 2$ (d) $4.9 \div 7$

   (e) $1.25 \div 5$ (f) $3.54 \div 3$ (g) $12.35 \div 5$ (h) $15.3 \div 9$

2 A plank of wood measuring 3.45 m is cut into 5 equal length pieces.
   How long is each piece?

3 Work out

   (a) $6.3 \div 2$ (b) $2.5 \div 2$ (c) $6.4 \div 5$ (d) $8.6 \div 4$

4 Find three matching pairs of division that give the same answer.
   Which is the odd one out?

   A $\boxed{5.8 \div 2}$ B $\boxed{1.44 \div 3}$ C $\boxed{3.4 \div 4}$ D $\boxed{14.5 \div 5}$

   E $\boxed{2.4 \div 5}$ F $\boxed{3.5 \div 2}$ G $\boxed{5.1 \div 6}$

5 A piece of cheese weighing 1.3 kg is cut into two pieces of equal weight.
   How heavy is each piece?

6 A plank of wood 6 metres long is cut into 4 equal lengths.
   How long is each piece?

7 15 kg from a sack of flour is divided into 4 equal bags.
   How much flour is in each bag?

8 Match each division with an answer from the hoop.
   You can use the numbers more than once.

   (a) $9 \div 2$ (b) $7 \div 4$ (c) $6.2 \div 5$

   (d) $5.25 \div 3$ (e) $10 \div 4$ (f) $18 \div 4$

   (g) $22.5 \div 5$ (h) $16 \div 5$ (i) $7.5 \div 3$

   (1.75   4.5   1.24   2.5   3.2)

9 Find the answers to these questions then use this code to change them to letters.
   Rearrange each set of letters to spell a fruit.

A	C	E	G	H	L	P	R	Y
1.5	1.25	0.2	0.5	1.7	0.3	0.4	2.5	0.82

   (a) $4.1 \div 5$ (b) $2.4 \div 6$ (c) $10 \div 8$ (d) $0.8 \div 2$
   $\quad\ \ 10 \div 4$ $\quad\ \ 2.7 \div 9$ $\quad\ \ 8.5 \div 5$ $\quad\ \ 7.5 \div 5$
   $\quad\ \ 5.1 \div 3$ $\quad\ \ 0.8 \div 4$ $\quad\ \ 3 \div 2$ $\quad\ \ 5 \div 2$
   $\quad\ \ 2.5 \div 2$ $\quad\ \ 2 \div 5$ $\quad\ \ 1.6 \div 4$ $\quad\ \ 4 \div 8$
   $\quad\ \ 1.6 \div 8$ $\quad\ \ 6 \div 4$ $\quad\ \ 1 \div 5$ $\quad\ \ 1.2 \div 6$
   $\quad\ \ 7.5 \div 3$

Section H

1 Tracy was making up orange squash in a 3 litre container.
  She poured in 0.45 litres of orange, and then filled the container with water.
  How much water did she add?

2 Find the areas of these rectangles.

(a)
6 cm
38 cm

(b)
2.6 m
7 m

3 This pie chart shows what transport people
  used to get to a town shopping centre.

  What percentage came by train?

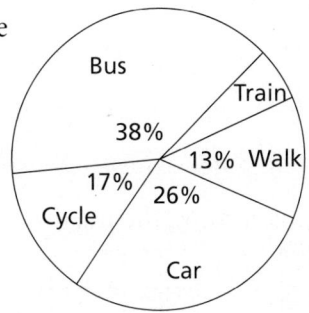

Bus
Train
38%
13% Walk
17%
26%
Cycle
Car

4 A one pound coin has a diameter of 2.2 cm.
  How long would a line of 8 one pound coins be?

5 The perimeter of a rectangle is the distance all the way round it.

  This rectangle has perimeter 3.4 metres and width 0.8 m.
  What is its length?

0.8 m

6 This square has perimeter 23 cm.
  How long is each side?

7 To find the mean of a set of numbers you add them all up,
  then divide by how many there are.

  These are the heights of some young trees. Find their mean height.

  3.2 m   4.1 m   5.6 m   4.7 m   5.2 m

8 To find the area of a triangle use the formula

  Area triangle = base × height ÷ 2

  Find the areas of these triangles.

(a)
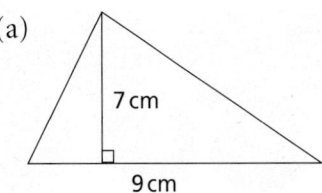
7 cm
9 cm

(b)
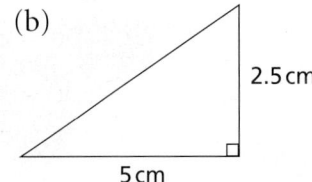
2.5 cm
5 cm

23

22 *Fractions*

Sections A, B and C

1 What fraction of each shape is shaded? (a) 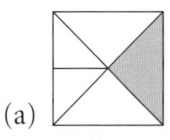 (b)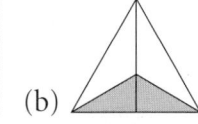

2 The diagrams show two equivalent fractions.

   What fractions are they?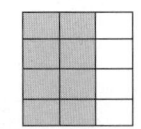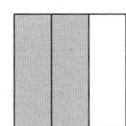

3 Copy these and find the missing numbers
   (a) $\frac{1}{2} = \frac{}{8}$
   (b) $\frac{1}{3} = \frac{}{12}$
   (c) $\frac{2}{5} = \frac{}{15}$
   (d) $\frac{3}{4} = \frac{}{16}$

4 Write each of these fractions in its simplest form.
   (a) $\frac{6}{10}$
   (b) $\frac{5}{15}$
   (c) $\frac{8}{12}$
   (d) $\frac{12}{20}$
   (e) $\frac{9}{21}$

5 Write in its simplest form, the fraction of the square that is shaded.

   (a) 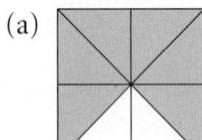 (b) 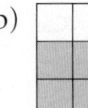 (c) 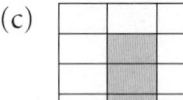 (d)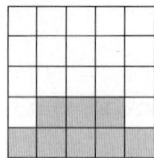

6 (a) Which fractions in this list are equivalent to $\frac{1}{3}$?

   $\frac{3}{9}$ $\frac{5}{20}$ $\frac{6}{24}$ $\frac{6}{18}$ $\frac{2}{6}$ $\frac{4}{16}$ $\frac{4}{20}$ $\frac{8}{24}$

   (b) Which fractions in the list above are equivalent to $\frac{1}{4}$?

Section D

In these questions write each fraction in its simplest form.

1 What fraction of this set of pens have their caps missing?

2 What fractions of these coins are showing 'Heads'?

3 Patrice took 24 photos on holiday.
16 of the photos were of people. The rest were of places.

What fraction of Patrice's photos were of people?

4 Six children in a class of 30 are left-handed.
What fraction of the class are left-handed?

5 What fraction of
 (a) WEDNESDAY is WED   (b) MANCHESTER is CHEST
 (c) BRONTOSAURUS is TO   (d) UNREPRESENTATIVE is SENT

6 Class 10R consists of 12 fourteen-year-olds and 20 fifteen-year-olds.
What fraction of the class are fifteen-year-olds?

7 Last season, Holby City football team played 24 matches.
They won 10, drew 8 and lost 6.
What fraction of their matches did they
 (a) win   (b) draw   (c) lose?

8 Josie keeps rabbits.
She has 12 white rabbits, 10 brown rabbits and 8 black rabbits.

What fraction of her rabbits are
 (a) white   (b) brown   (c) black

Section E

1 Work out
 (a) $\frac{1}{3}$ of 18   (b) $\frac{3}{4}$ of 16   (c) $\frac{1}{5}$ of 35   (d) $\frac{3}{5}$ of 35   (e) $\frac{2}{3}$ of 36

2 Work out
 (a) $\frac{5}{8}$ of 40   (b) $\frac{3}{5}$ of 45   (c) $\frac{3}{8}$ of 160   (d) $\frac{4}{5}$ of 250   (e) $\frac{5}{6}$ of 300

23 Circle facts

Section B

1 Work out roughly the circumference of each of these tins.

    (a) diameter 5 cm   (b) diameter 12 cm   (c) diameter 16 cm

2 This wedding cake has three layers.
A ribbon goes round the outside of each layer.

    (a) The diameter of the bottom cake is 30 cm.
        Roughly how long will the ribbon be?

    (b) The diameter of the middle layer is 20 cm.
        Roughly how long will this ribbon be?

    (c) The diameter of the top cake is 15 cm.
        Roughly how long will the top ribbon be?

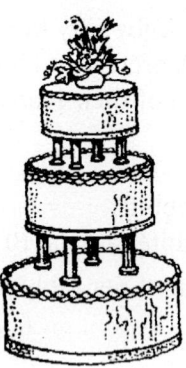

3 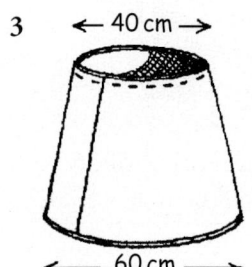 This lampshade needs a piece of tape sewn round
the top and another piece at the bottom.
Roughly how much tape will be needed altogether?
(Show your working clearly)

4 This traffic island is a circle 15 metres across.
New kerbstones are being laid around its circumference.
Each kerbstone is $\frac{1}{2}$ metre long.

Roughly how many will be needed?

26

5 Cosima is putting tape round this roll of carpet.
The diameter of the roll is about 25 cm.

(a) Roughly what length of tape will she need to
go round the carpet once?

(b) She puts three pieces of tape round the roll.
About how much tape does she need altogether?

6

Ferris Wheel 1893

This is a picture of the first Big Wheel ever built.
It was 40 m from the centre to the 'cars' on
the outside.

(a) About how high from the ground
was the top car?

(b) Roughly what was the circumference
of the wheel?

Section C

1 Mary and Paul are measuring some trees.
They measure the circumference of each tree.
Then they work out roughly the diameter
and radius of each tree.

Copy and complete the table below
for some trees they measured.

Type of tree	Circumference	Diameter	Radius
Oak	120 cm		
Silver Birch	60 cm		
Horse chestnut	150 cm		
Yew	210 cm		
Beech	75 cm		

Section D

Use the π button on your calculator when doing these questions.
Give each answer to the nearest 0.1 cm.

1 Work out the circumference of each of these circles.

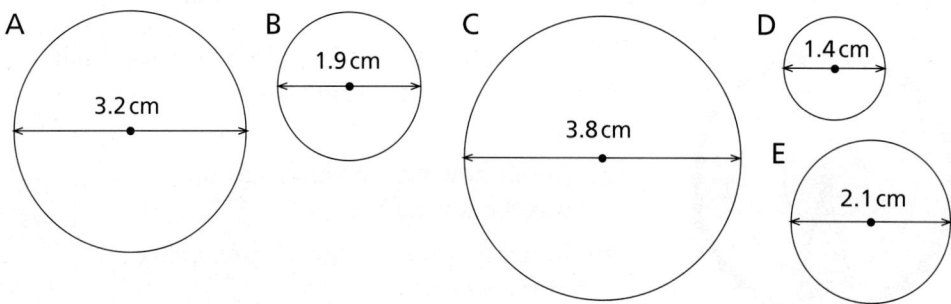

A 3.2 cm
B 1.9 cm
C 3.8 cm
D 1.4 cm
E 2.1 cm

2 Measure the radius of each of
these circles, and write it down.
Then work out the circumference
of each circle.

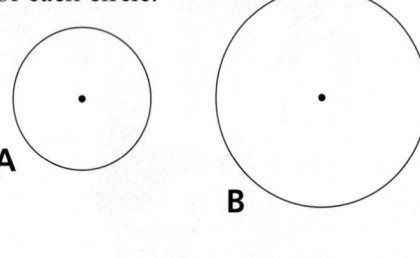
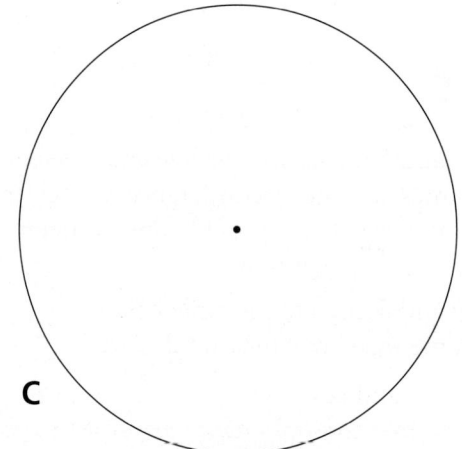

A

B

C

3 Harry cuts a piece of plastic 80 cm long to
go round the circular end of a lampshade.
The diameter of the lampshade is 26 cm.

Is the plastic long enough?
Explain your answer.

26 cm

4 Bill designs labels to go round cans of food.

How tall and how long will a label
need to be to go round this tin?

8.5 cm

6 cm

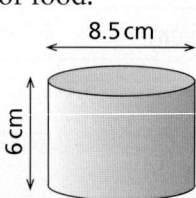

28

Mixed practice 4

1 This chart shows the ages of people using a swimming pool on a Sunday morning.

   (a) How many people aged 40 or over used the pool?

   (b) How many people altogether used the pool?

   (c) What is the modal age group of people using the pool this Sunday morning?

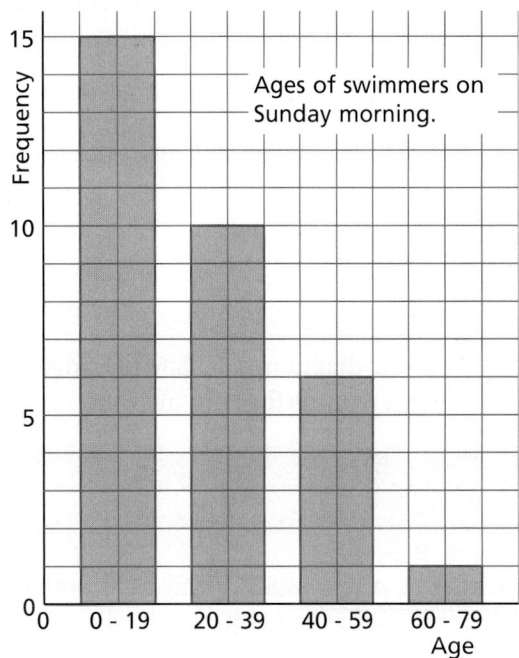

Ages of swimmers on Sunday morning.

2 The table below shows the ages of people using the swimming pool on a Monday morning.

Weight in kg	65	66	68	46	63	66	42	60	65	56
Ages of swimmers	45	61	59	48	59	59	62	68	69	53

   (a) Copy and complete this stem and leaf table. The first two ages have been put into the table.

   (b) Write out the table again, putting the leaves in order.

   (c) What is the median age of these swimmers?

   (d) What is the range of the ages of these swimmers?

   (e) Write down what you notice about the ages of people using the pool on Monday morning compared with Sunday morning.

```
0 |
1 |
2 |
3 |
4 |
5 |
6 | 5 6
Stem = 10 years
```

3 Copy the table on the right. Put ticks in the correct columns for the numbers in each row.

	is a factor of 24	is a multiple of 8	is a prime number	is a square number	is a cube number
8					
16					
27					
32					
36					
64					

4 Do these in your head, without writing down any working.

   (a) 22×4 (b) 34×5 (c) 13×4 (d) 23×5 (e) $340 \div 5$

   (f) $72 \div 4$ (g) 5×83 (h) $260 \div 5$ (i) $180 \div 4$ (j) $330 \div 5$

5 Without using a calculator, work out each of these. Show your working clearly.

   (a) $427 + 126$ (b) $531 - 207$ (c) 54×8 (d) $2412 \div 6$

   (e) $4.3 + 2.86$ (f) $4 - 0.54$ (g) 3.75×4 (h) $6.15 \div 5$

6 Calculate the missing angles in these diagrams.

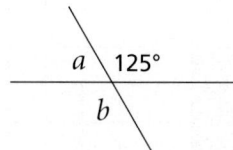 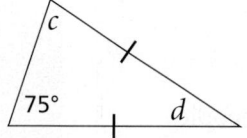 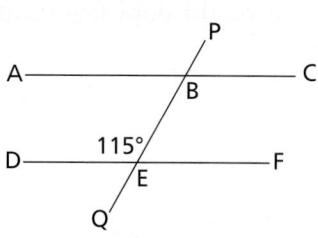

7 In the diagram, calculate the angle PBC.
Give reasons for your answer.

8 (a) A tree in Sicily had a circumference of 57 metres.
What, roughly, was its diameter?

   (b) A bicycle wheel has a radius of 35 centimetres.
About how many times would it go round
if the bicycle travelled 21 metres?

9 Write, in its simplest form, the fraction of each rectangle that is shaded.

   (a) 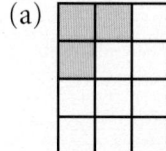   (b) 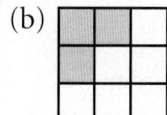   (c) 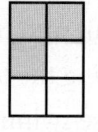   (d) 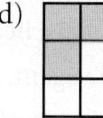   (e)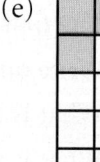

10 (a) Salvador has 20 rabbits. 12 of them are male.
What fraction of Salvador's rabbits are male?
Write the fraction in its simplest form.

   (b) Salvador also has 24 piglets.
$\frac{3}{8}$ of them are less than 6 months old.
How many of his piglets are less than 6 months old?

   (c) He also has 12 male pigeons and 18 female pigeons.
What fraction of Salvador's pigeons are female?
Write the fraction in its simplest form.

24 *Fractions, decimals and percentages*

Sections A, B and C

1 Copy and complete this table.

Fraction	Decimal	Percentage
$\frac{3}{4}$		
	0.6	
$\frac{7}{10}$		
		40%
	0.9	

2 Write each of these lists in order of size, smallest first.

(a) $\frac{4}{5}$, 0.9, $\frac{3}{4}$, 70%, $\frac{1}{2}$

(b) $\frac{3}{5}$, $\frac{1}{2}$, $\frac{3}{4}$, 40%, 0.8

Sections E and F

1 Work out

(a) 10% of £40
(b) 20% of £40
(c) 70% of £40

(d) 30% of £20
(e) 40% of £50
(f) 60% of £40

2 A restaurant adds 10% to the cost of a meal, for service.
Jasons meal costs £15.

(a) How much does the restaurant add on for service?

(b) How much does Jason pay altogether?

3 20% of an orange's weight is peel.
If an orange weighs 80 g, how much does its peel weigh?

4 Copy and complete this table.

Fraction	Decimal	Percentage
$\frac{81}{100}$		
	0.01	
		87%
		6%

5 Write each of these lists in order of size , smallest first.

(a) 0.27, $\frac{1}{4}$, $\frac{1}{5}$, 30%, 0.4

(b) 0.8, $\frac{3}{4}$, $\frac{3}{5}$, 0.67, $\frac{9}{10}$

(c) 25%, $\frac{1}{5}$, $\frac{1}{2}$, 0.06, 0.1

(d) $\frac{4}{5}$, 0.9, $\frac{3}{4}$, 0.85, 60%

25 Negative numbers

Section A

1 Work these out.

   (a) $^-3 + 9$ (b) $^-4 + 1$ (c) $^-11 + 3$ (d) $^-6 + 8$ (e) $^-5 - 2$

   (f) $^-1 - 5$ (g) $2 - 5$ (h) $4 - 7$ (i) $^-3 - 2$ (j) $^-5 + 3$

2 Work these out.

   (a) $^-3 + 8$ (b) $^-10 + 2$ (c) $2 - 9$ (d) $^-3 - 10$ (e) $2 - 12$

   (f) $^-4 - 5$ (g) $^-3 + 16$ (h) $^-20 + 5$ (i) $7 - 14$ (j) $^-12 - 4$

Section B

1 Work these out.

   (a) $3 + {}^-1$ (b) $7 + {}^-2$ (c) $^-4 + {}^-2$ (d) $10 + {}^-3$ (e) $^-1 + 9$

   (f) $^-6 + {}^-2$ (g) $^-3 + {}^-7$ (h) $^-2 + 11$ (i) $^-8 + {}^-4$ (j) $3 + {}^-9$

2 From the numbers in the loop, find two numbers that add up to

   $\boxed{^-6 \quad ^-4 \quad 2 \quad 5}$

   (a) $^-10$ (b) $^-1$ (c) 1 (d) $^-4$ (e) $^-2$

3 From the numbers in the loop, find two numbers that add up to

   $\boxed{^-10 \quad ^-6 \quad 3 \quad 5}$

   (a) $^-7$ (b) $^-3$ (c) $^-5$ (d) $^-1$ (e) $^-16$

Section C

1 Work these out.

   (a) $6 - {}^-2$ (b) $5 - {}^-1$ (c) $10 - {}^-3$ (d) $4 - {}^-2$ (e) $8 - {}^-6$

2 Work these out.

   (a) $^-4 - {}^-1$ (b) $^-2 - {}^-5$ (c) $^-3 - {}^-8$ (d) $3 - {}^-4$ (e) $1 - {}^-1$

3 Work these out.

   (a) $^-3 + 9$ (b) $^-3 - 9$ (c) $^-3 + {}^-9$ (d) $^-3 - {}^-9$

   (e) $^-2 + {}^-7$ (f) $^-2 - {}^-7$ (g) $2 - 7$ (h) $2 - {}^-7$

   (i) $4 - 8$ (j) $^-4 - 8$ (k) $^-4 - {}^-8$ (l) $4 - {}^-8$

Section D

1 Work these out

(a) $^-3 \times 5$ (b) $^-4 \times 7$ (c) $^-2 \times 8$ (d) $4 \times ^-3$ (e) $5 \times ^-5$

2 Work out the answers to the questions below.
Use the code to change them to letters.
Re-arrange the letters to make some countries.

A	E	I	L	N	P	R	S	T	U	Y
$^-20$	$^-5$	1	$^-8$	$^-3$	$^-18$	$^-6$	5	$^-15$	3	$^-2$

(a) $6 \times ^-3$
$1 - ^-2$
$^-2 \times 3$
$1 - 6$

(b) $^-5 - ^-2$
$3 + ^-2$
$^-2 + 7$
$2 \times ^-9$
$^-5 \times 4$

(c) $^-8 - ^-6$
$2 \times ^-10$
$^-4 \times 2$
$^-8 - 7$
$^-3 - ^-4$

(d) $3 - ^-2$
$0 - ^-3$
$3 \times ^-2$
$^-8 + 9$
$^-12 + ^-8$
$8 + ^-3$

Section E

1 Work out the value of

(a) $3a$ when $a = ^-7$ (b) $4b$ when $b = ^-4$ (c) $10c$ when $c = ^-6$

(d) $2d + 1$ when $d = ^-4$ (e) $3e + 2$ when $e = ^-5$ (f) $4f + 1$ when $f = ^-5$

2 Work out the value of

(a) $4p - 2$ when $p = ^-3$ (b) $2q + 7$ when $q = ^-3$ (c) $3r = 1$ when $r = ^-4$

(d) $2s - 6$ when $s = ^-5$ (e) $3t + 6$ when $t = ^-3$ (f) $4u - 3$ when $u = ^-10$

26 *Areas of parallelograms*

Sections B and C

1 Find the area of each of these parallelograms.

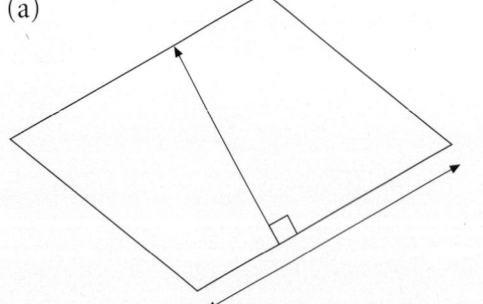

(a)

(b)

(c)

(d)

(e)

(f)

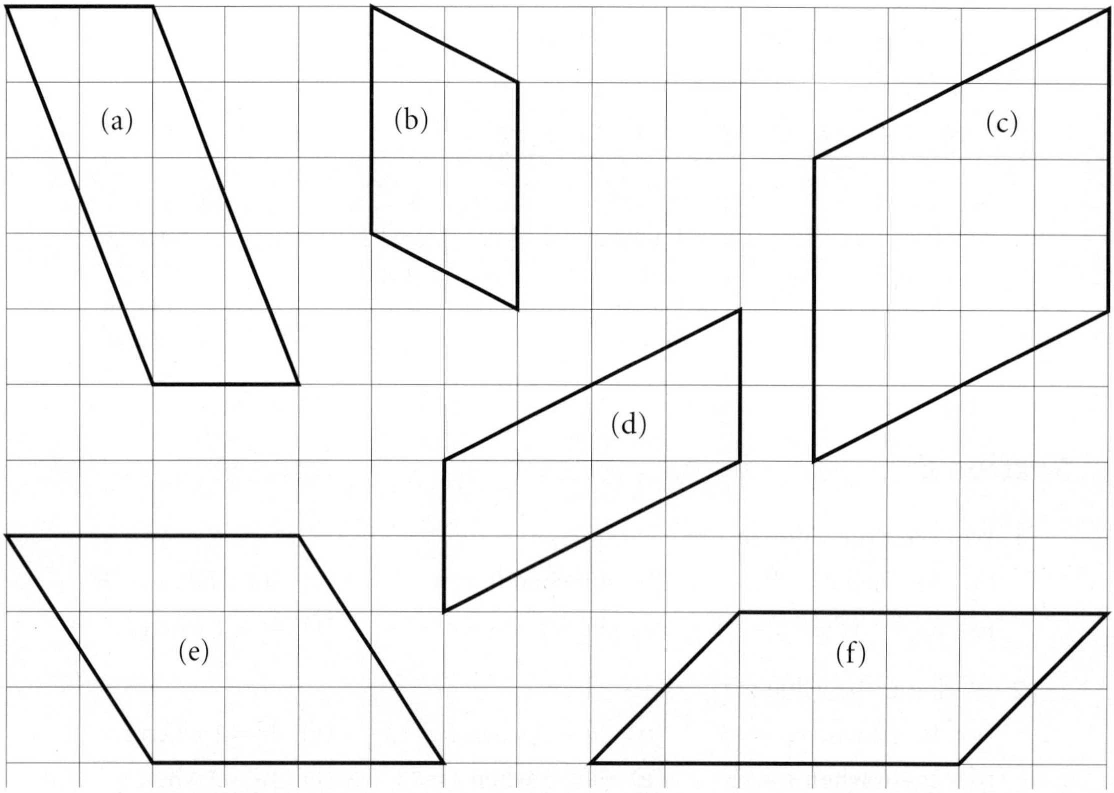

2 On centimetre squared paper, draw:

(a) three different parallelograms each with an area of $12\,\text{cm}^2$

(b) three different parallelograms each with an area of $15\,\text{cm}^2$

3 Find the areas of these parallelograms by measuring the base and the height shown.

(a)

(b)

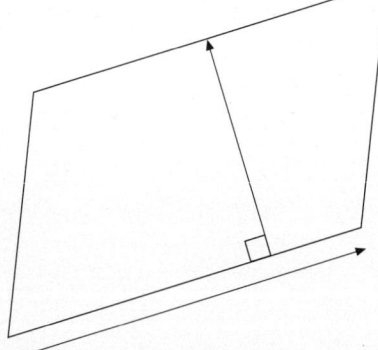

4 Find the areas of the parallelograms,

(a)
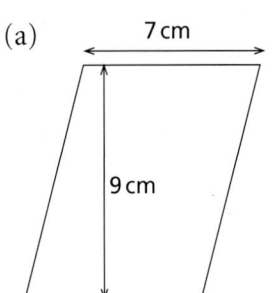
7 cm
9 cm

(b)
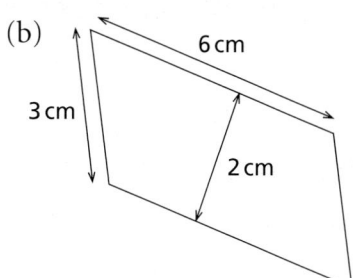
6 cm
3 cm
2 cm

(c)
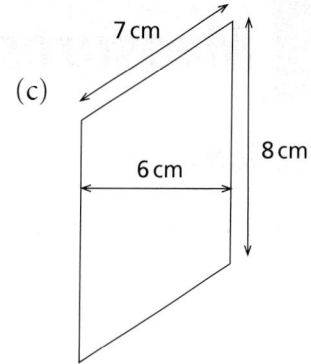
7 cm
8 cm
6 cm

(d)
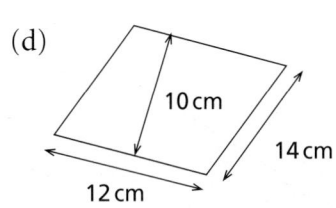
10 cm
14 cm
12 cm

(e)
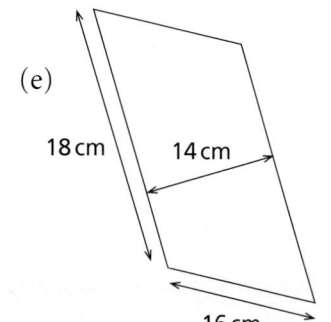
18 cm
14 cm
16 cm

(f)
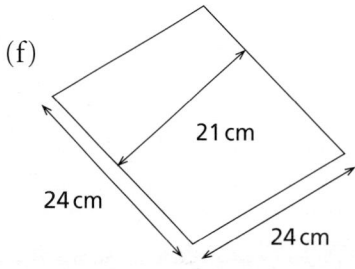
21 cm
24 cm
24 cm

Section D

1 Find the areas of these parallelograms.

(a)
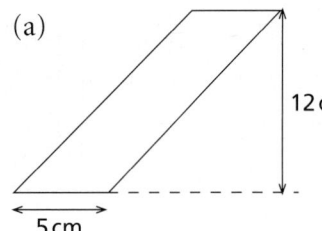
12 cm
5 cm

(b)
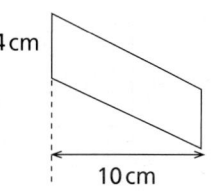
4 cm
10 cm

(c)
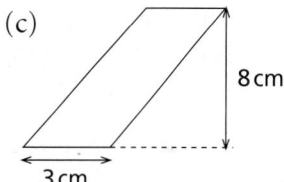
8 cm
3 cm

(d)
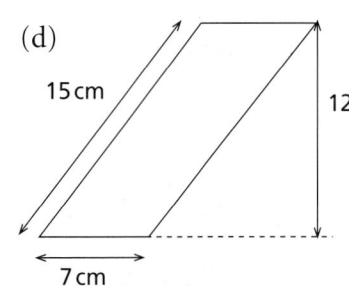
15 cm
12 cm
7 cm

(e)
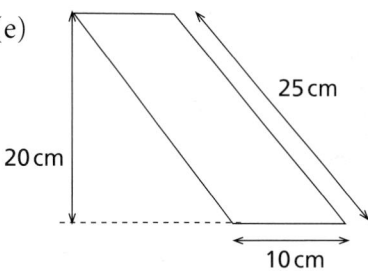
25 cm
20 cm
10 cm

(f)
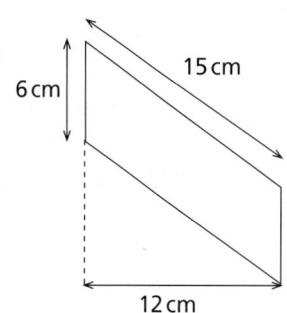
15 cm
6 cm
12 cm

27 **Measures**

Section A

1 The length of this drinking straw is 5 cm.

Estimate the length of these drinking straws.

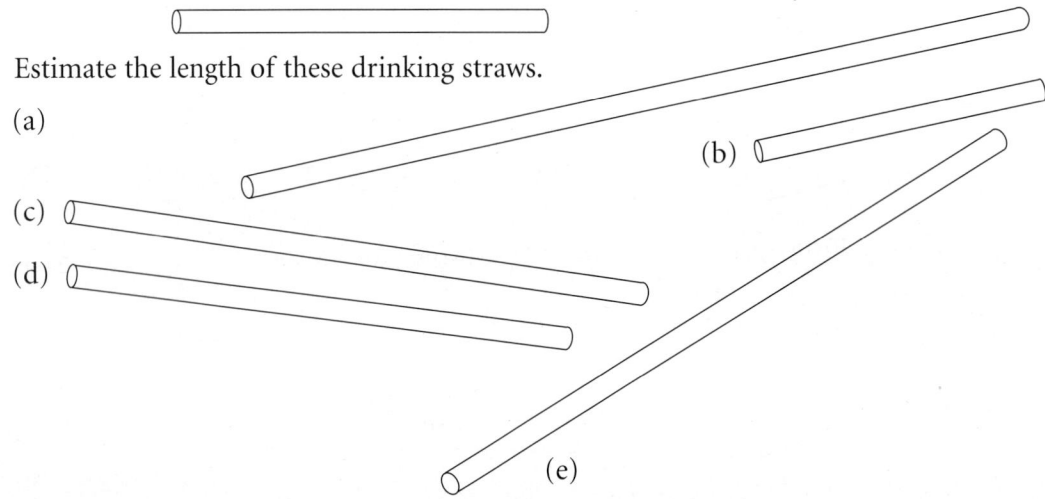

(a)

(b)

(c)

(d)

(e)

2 The diagram shows the plan of a tennis court.

The tennis court is 10 m wide. Estimate it's length.

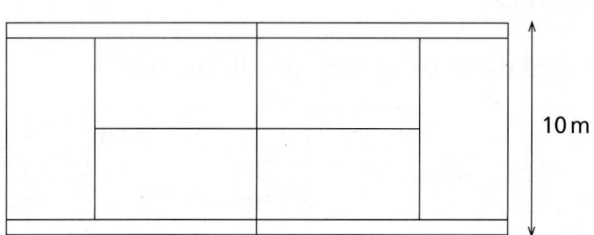

10 m

3 The height of the adult hippopotamus in this picture is 1.5 m.

Estimate the height of:

(a) the elephant

(b) the hyaena

(c) the giraffe

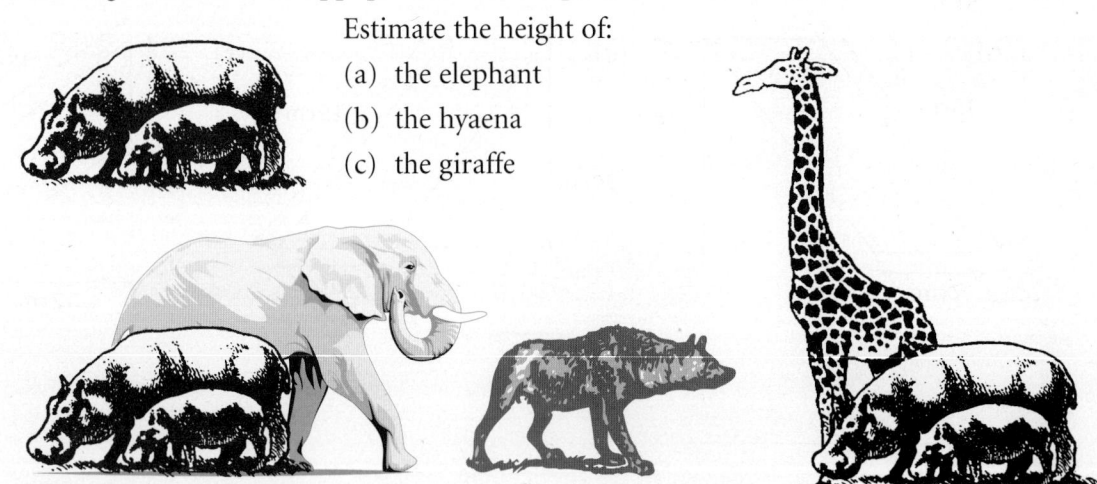

Section B

1 This dial is from a set of scales measuring in grams.

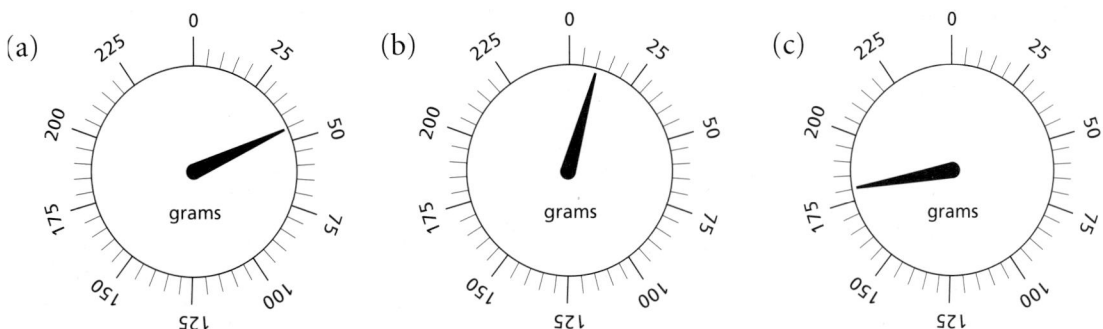

(a)   (b)   (c)

What weight is indicated by the pointer in these diagrams?

2 This is the dial from some bathroom scales.

It measures weight in kilograms.

(a) What does each small division represent?

(b) The 'pointer' is the vertical line.

What weight is shown on the scales?

(i) 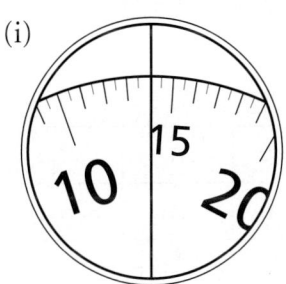   (ii) 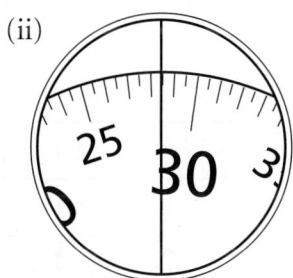   (iii)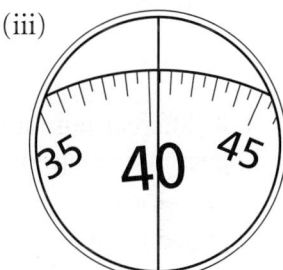

3 These scales are from a spring balance
used in science lessons.

(a) What does each small division represent?

(b What weight is shown by each scale?

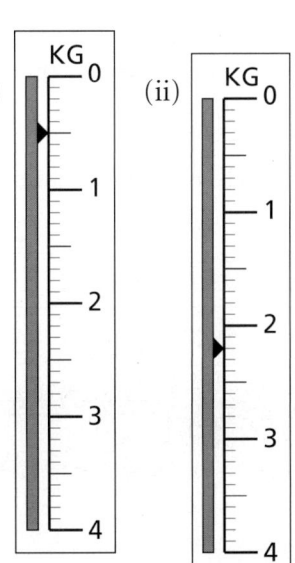

(i)   (ii)

28 *Gathering like terms*

Section A

1 Write an expression for the perimeter of each shape.
Write each expression as simply as possible.

(a)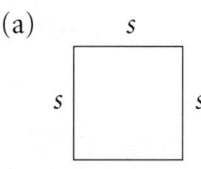

(b)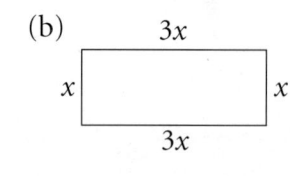

(c)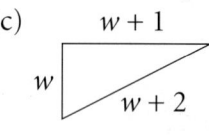

(d)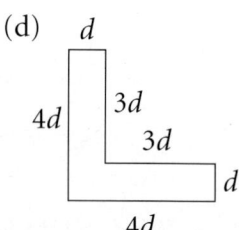

2 Simplify each of these expressions

  (a) $x + x + x$
       (b) $2e + 4 + e - 1$
       (c) $j + 2j + j - 6$

  (d) $m + 2 + 2m - 3$
   (e) $w + 2 - 2w + 5w + 8$
  (f) $100w + 10w + 7$

3 A number track goes up in steps of 2. You can use algebra to show it, using x.
Write down the three missing expressions.

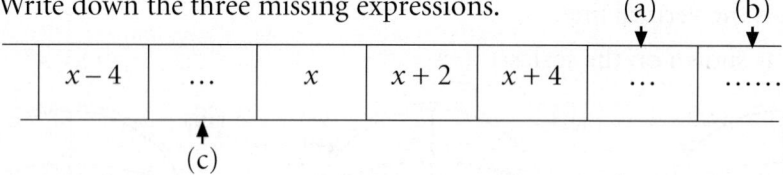

4 A different number track
goes up in steps of 3.

Two of these tracks are
joined together.

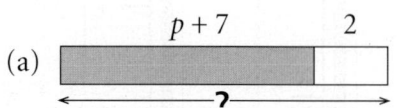
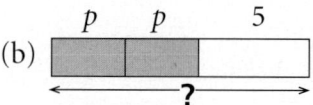

  (a) The sum of column A is $x + x$. Write this as simply as possible

  (b) Write expressions for the sum of

     (i) Column B   (ii) Column C   (iii) Column D

5 Look at these rods. Write an expression for each length marked **?**

(a)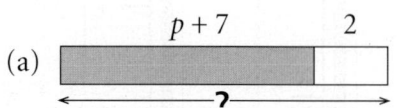

(b)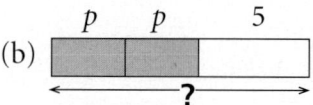

(c)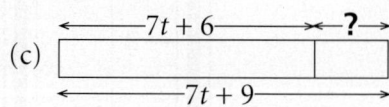

Sections B and C

1 Work out and simplify an expression for the perimeter of each of these.

(a)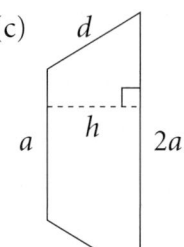

(b)

(c)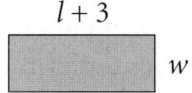

(d)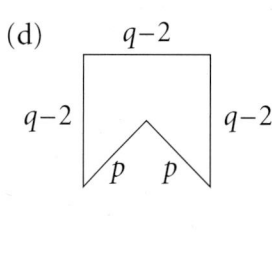

2 (a) Work out and simplify an expression
for the perimeter of this rectangular tile.

$l + 3$

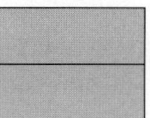

(b) Two of the tiles are put together as shown.

Write an expression for the perimeter of this shape.
Write your answer as simply as possible.

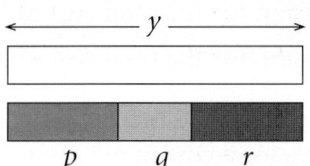

3 Look at these rods.
The two rods are the same length.

Decide whether each of these
equations is true (**T**) or false (**F**).

(a) $y = p + q + r$ (b) $y = q + r + p$ (c) $q + r = y - p$

(d) $y - q = p - r$ (e) $y - q = p + r$ (f) $y = 3p$

4 Look at these rods.
Write an expression for each length marked **?**

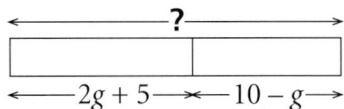

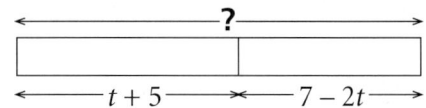

5 Sketch each diagram and write an
expression for the missing angles.

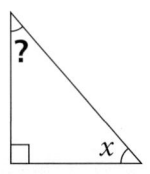

right-angled
triangle

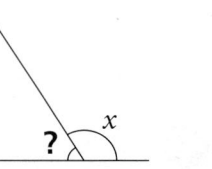

isosceles
triangle

Section D

1 A rectangle has width w cm.
The length is 4 cm more than the width.

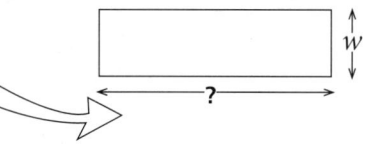

(a) Write down an expression for the length of the rectangle.

(b) If P is the perimeter of the rectangle,
write down a formula for P in terms of w.

2 Andrew has x ping-pong balls in his box.
His mum buys him four more balls.

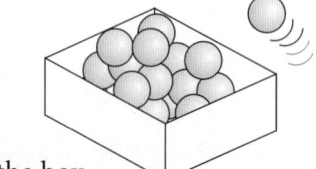

(a) Write an expression for the total
number of ping-pong balls he has now.

(b) His younger brother 'borrows' 6 balls from the box.
Write an expression for the number of balls in the box now.

3 Paper cups cost 12p each.

(a) Write an expression for the cost of n paper cups.

(b) Paper plates cost p pence each.
Write an expression for the total cost of 1 paper cup and 1 plate.

(c) Write an expression for the total cost of 3 paper cups and 3 plates.

4 (a) I have a roll of material 10 metres long.
I cut off x metres for a costume.
How much material is left?

(b) I have another roll of material w metres long.
I need to cut off material for 2 costumes
(each needs x metres).
How much material is left?

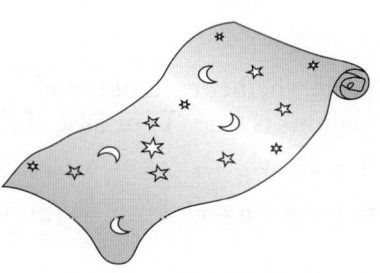

5 Here are two sets of algebra cards.
Match each card on the left with one from the right.

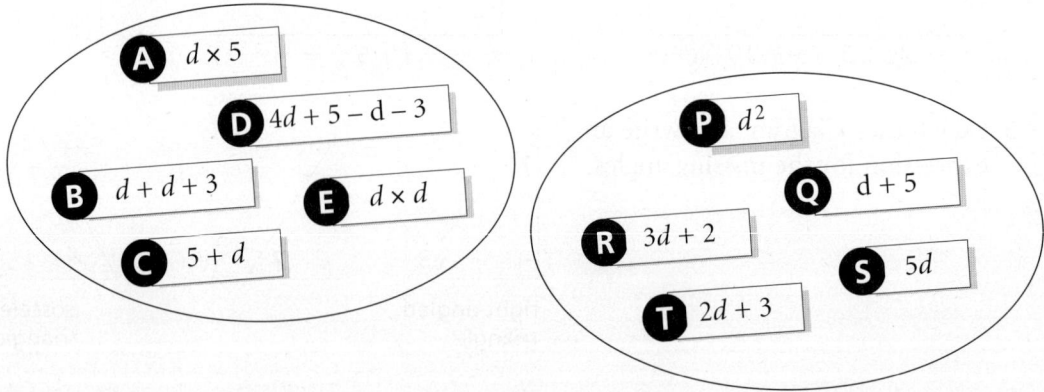

A $d \times 5$

D $4d + 5 - d - 3$

B $d + d + 3$

E $d \times d$

C $5 + d$

P d^2

Q $d + 5$

R $3d + 2$

S $5d$

T $2d + 3$

30 *Written calculation 2*

Section A

1 Work these out

   (a) 14 × 18   (b) 23 × 41   (c) 42 × 35   (d) 37 × 27

2 Work these out

   (a) 56 × 73   (b) 74 × 82   (c) 68 × 62   (d) 83 × 91

3 Work these out

   (a) 175 × 21   (b) 432 × 29   (c) 721 × 92   (d) 63 × 231

4 (a) The diagram shows a block of seats in a football stadium.

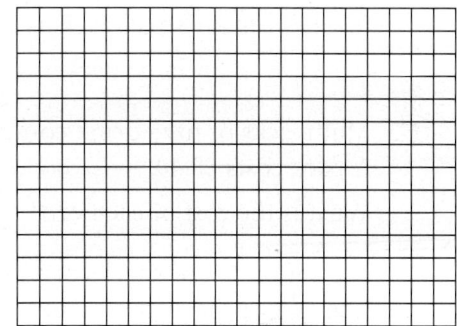

      (i) How many seats are there in the block?

      (ii) There are 16 of these blocks of seats in the West Stand. How many seats are there in the West Stand?

  (b) The supporters club took 18 coaches of fans to the away match. Each coach carried 41 supporters.

      (i) Do an approximate calculation to estimate how many supporters travelled by coach.

      (ii) Calculate the exact number of supporters who travelled by coach

  (c) The price of the tickets for the match was £24 each. What was the cost of 328 tickets?

  (d) The souvenir shop at the ground sold 18 car stickers at 55p each. How much money did they get for the car stickers?

  (e) The food kiosk sold 123 bars of chocolate costing 38p each and 247 fizzy drinks costing 44p each.

      (i) How much money did they get for the bars of chocolate?

      (ii) How much money did they get for the fizzy drinks?

Section B

1 Work these out

   (a) $490 \div 14$ (b) $576 \div 18$ (c) $748 \div 22$ (d) $384 \div 24$

2 Work these out

   (a) $713 \div 23$ (b) $912 \div 48$ (c) $901 \div 53$ (d) $999 \div 37$

3 Work these out

   (a) $1794 \div 26$ (b) $1998 \div 37$ (c) $6080 \div 64$ (d) $1508 \div 58$

4 A seed merchant packed his tomato seeds into small and large packets.

He put 18 seeds into each small packet and 32 seeds into each large packet. He had 1440 seeds.

(a) How many small packets could he make?

(b) How many large packets can he make?

5 Multipacks of fizzy drink contain 24 cans.
A pack costs £8.40.

What is the cost of each can?

6 A giant packet of snacks contains 12 small packets.
The cost of the giant packet is £3.24.

What is the cost of a small packet of snacks?

7 Oranges are sold in bags of 15 for £2.85 or bags of 12 for £2.52

(a) What is the price of an orange from each bag?

(b) Are the oranges in the bigger bag cheaper and, if so, by how much each?

8 James uses pieces of string 15cm long to tie up his rubbish bags.
He has some string 840cm long.

How many rubbish bag ties can he cut from the string?

9 648 chairs are arranged in the school hall for assembly each week.
When form 9RX arranged the chairs they put 24 chairs in each row.
Form 9SW decided to put 18 chairs in each row when it was their turn to arrange the chairs.

How many rows of chairs were arranged by

(a) 9RX   (b) 9SW?

10 Find the marked lengths for these rectangles.

(a)

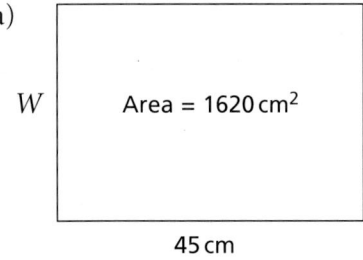

W   Area = 1620 cm²

45 cm

(b)

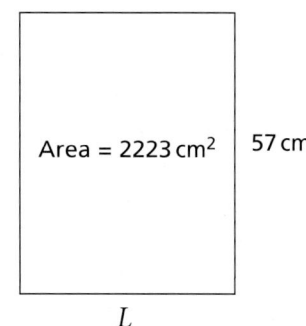

Area = 2223 cm²   57 cm

L

11 Copy and complete this division cross-number

Across Clues

(1) 544 ÷ 34

(3) 368 ÷ 16

(4) 1107 ÷ 27

(5) 1300 ÷ 25

Down Clues

(1) 748 ÷ 44

(2) 345 ÷ 15

(3) 399 ÷ 19

(4) 924 ÷ 22

12 Margaret has 156 CDs.
She keeps them in racks.
Each rack holds a maximum of 24 CDs.

How many racks does she need?

13 A large bottle of drink holds 2500 ml of orange juice.

(a) How many glasses each holding 85ml can be filled from the bottle?

(b) How much orange juice is left over?

14 A bag of lawn feed weighs 2 kg
The instructions suggest it is spread at the rate of 35 g per square metre.

How many square metres will the bag cover?

Section C

1 4 and 5 are called a pair of factors of 20 because 4 × 5 = 20
31 is one factor of a pair of factors of 1457

Find the other factor of the pair?

2 Jamie's car travels 11 miles on every litre of petrol.
Last month he drove 396 miles.

(a) How many litres of petrol did he use?

(b) If a litre of petrol costs 92p, how much did Jamie spend on petrol last month?

3 To celebrate leaving school, Sam's class went to a restaurant for a meal.
They decided to split the cost equally between them.
The meal cost £414.
There were 18 people in Sam's class.

How much did they each pay?

4 A bathroom wall is 360 cm long and 270 cm high.

It is covered in rows of square tiles each measuring 15 cm by 15 cm.

(a) How many rows of tiles are there?

(b) How many tiles are there in each row?

(c) How many tiles are there on the wall?

(d) Each tile cost 36p.
What was the total cost of the tiles?

31 Connections

Section A

1 Two film critics watched some films at a film festival and gave each film a score out of 10.
The scores were:

	Film									
	A	B	C	D	E	F	G	H	I	J
Critic 1	6	8	2	4	4	9	7	6	7	10
Critic 2	6	7	2	6	5	10	4	8	7	8

(a) Which film did critic 1 like the least?

(b) Which film was critic 2's favourite?

(c) Draw a scatter diagram on squared paper with both scales going from 0 to 10 marking each film with a cross.

(d) Do you think that the film critics like similar films?
Give a reason for your answer.

2 The table shows the measurements in centimetres of the length and width of ten leaves taken from a tree.

	Leaves									
	A	B	C	D	E	F	G	H	I	J
Width	1.3	2.9	2.7	1.9	3.1	2.3	2.7	3.3	2.5	1.8
Length	3.2	6.6	4.8	4.0	6.5	4.9	5.5	7.0	5.0	3.0

(a) What is the length of the longest leaf?

(b) Is the longest leaf also the widest leaf?

(c) Plot the measurements on a scatter diagram with axes marked like this.

(d) What does the scatter diagram tell you about the relationship between the lengths and the widths of the leaves?

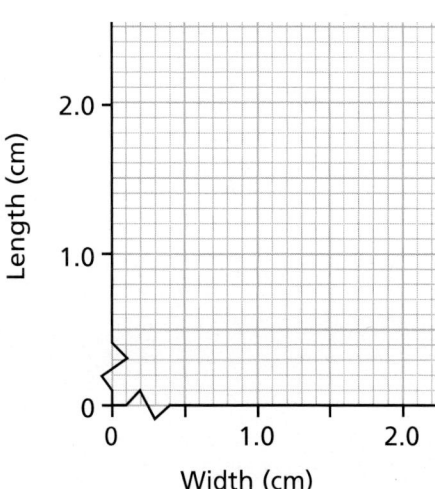

Section B

1 A group of students took a mathematics examination in year 6 and another mathematics examination in year 7.

Their results were as follows:

	Student											
	A	B	C	D	E	F	G	H	I	J	K	L
Year 6 (%)	93	84	62	76	55	98	66	51	88	61	58	79
Year 7 (%)	90	81	75	50	50	98	69	43	75	50	63	78

(a) Draw a scatter diagram on graph paper using these scales.

(b) What type of correlation does the graph show?

(c) What is the connection between the students results in year 6 and year 7?

(d) Which student performed much better in year 6 than year 7?

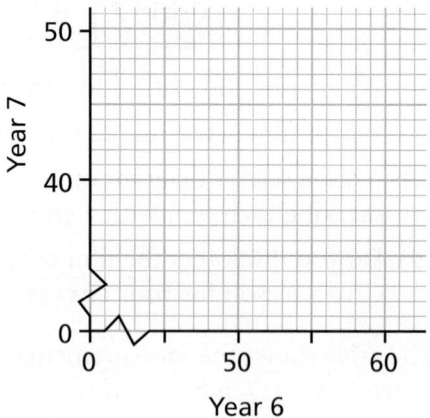

2 The table gives information taken from advertisements about the age of 9 cars and their sale price. All the cars are the same model.

Age	3	9	7	6	13	10	5	2
Value	£6000	£1000	£2000	£2500	£500	£2000	£3500	£6000

(a) Draw a scatter diagram on graph paper using these scales.

(b) What type of correlation does the graph show?

(c) Describe the relationship between the age and the value of the cars.

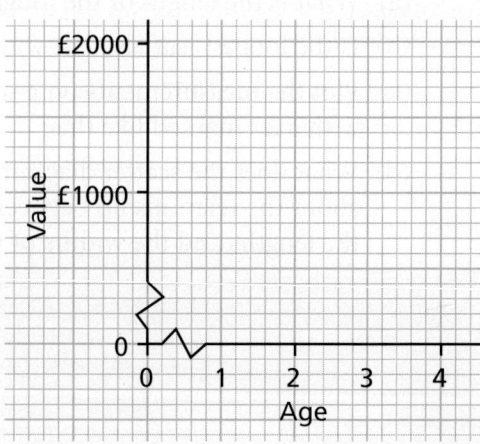

Section C

1 An 'Outdoor Life' magazine tested some different makes of walking shoes.
   They gave them a score based on comfort, water resisitance and durability.

	Shoes							
Cost (£)	30	60	65	50	70	80	70	50
Score	55	66	73	66	80	87	75	69

(a) Show the information on a scatter diagram
     using these scales.

(b) Describe the correlation between the cost
     and the score.

(c) Draw the line of best fit.

(d) A pair of walking shoes cost £40.
     Use your graph to estimate a score.

(e) Another pair of walking shoes scored 92 marks.
     How much would you expect them to cost?

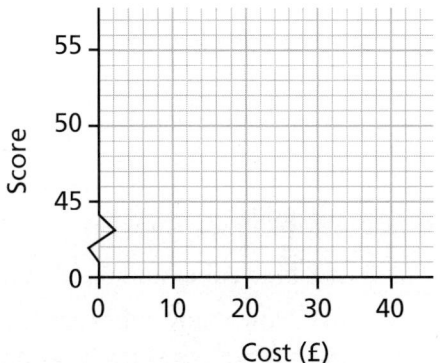

2 A football fan wanted to investigate any connection between the number of
   points obtained by a football team and the number of goals that were scored
   against them.

   He looked at 12 teams in the league.

Points	69	65	52	44	38	55	50	44	53	36	33	24
Goals against	25	22	34	32	32	31	28	40	23	39	46	46

(a) Use the data to draw a scatter graph on graph paper using these scales.

(b) Draw a line of best fit.

(c) What does the graph tell you about the
     connection betweeen the points
     and goals against?

(d) Another team obtained 48 points.
     Estimate how many goals were scored
     against them.

(e) If a team had 50 goals scored against them,
     estimate how points they obtained

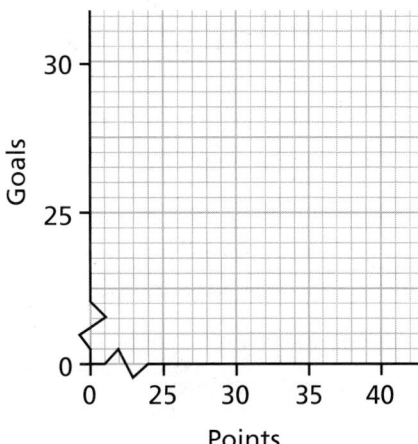

47

32 *Chance*

Section A

1 On a cube 3 faces are painted green, 2 are painted red and the other face is painted white.
The cube is rolled and the colour of the top face noted.
What is the probability that the top face is:

(a) red   (b) white   (c) green   (d) not red   (e) yellow

2 Seven balls numbered 1 to 7 are placed in a lottery machine.
One ball is released.
What is the probability that the number on it is:

(a) a 5   (b) not a 6   (c) even

(d) odd   (e) smaller than 3   (f) an 8?

3 If 8 blue cubes and 5 red cubes are put in a box

(a) How many cubes are there in the box?

One of the cubes is taken out of the box without looking.
What is the probability that the cube is:

(b) blue   (c) red?

4 A bag contains 3 dark chocolate cookies, 5 raisin cookies and
4 white chocolate cookies.
A cookie is taken from the bag without looking.

What is the probability that the cookie is

(a) a dark chocolate cookie   (b) a raisin cookie

(c) a chocolate cookie?

5 These 7 cards have pictures of shapes on one side.
The cards are placed picture side down and shuffled.

A card is turned over at random.

What is the probability that it is:

(a) a black picture

(b) a shape with 4 sides

(c) not a white picture

(d) a circle

(e) not a square.

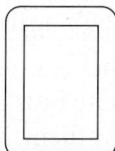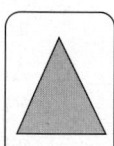

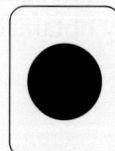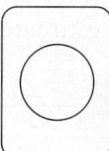

Section B

1 Jim has 3 shirts and 2 ties which he wears to work.
He has a white shirt, a grey shirt and a pink shirt.

One of his ties is red and the other is mauve.

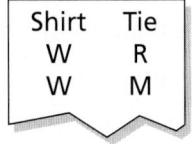

(a) Copy and complete this list of all the possible
combinations of shirt and tie that Jim could wear to work.

(b) How many different possible combinations are there?

He also has 2 pairs of trousers; a black pair and a tan pair.

(c) Make a list of all the different possible combinations of shirt,
tie and trousers that Jim could wear to work.

(d) How many different outfits are there?

2 For breakfast Jim eats cereals, toast and marmalade and has a drink of fruit juice.
He can choose from 4 different cereals:
cornflakes, wheatbiscuits, rice crunchies or porridge.

He can choose between orange or lemon marmalade.

(a) Copy and complete this list of all the possible
combinations of cereals and types of marmalade
he can eat for breakfast.

Cereal	Marmalade
C	O
C	L
W	O

He drinks either pineapple or grapefruit juice.

(b) List all the complete breakfasts he could chose.

(c) How many different breakfasts are there altogether?

3 Jim buys a sandwich for lunch from
Jane's Sandwich Bar.

> **JANE'S SANDWICH BAR**
> **we sell over
> 20 different sandwiches**

The sandwich bar uses 4 different kinds of bread –
white, brown, granary and harvestgrain.

The bread is spread with butter or margarine.

(a) Make a list of all the different possible combinations of bread and spread.

For fillings, Jane uses ham, cheese or egg.

(b) Make a list of all the different possible sandwich combinations.

(c) How many different sandwiches are available?

(d) Did the sandwich bar sell more than 20 different kinds?

Section C

1 The arrows on these spinners are spun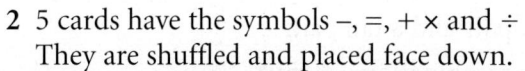

   (a) List all the possible outcomes

   (b) What is the probability of getting Black and 2

   (c) What is the probability of getting grey and an even number

   (d) What is the probability of getting a 3 and a colour which is not white?

2 5 cards have the symbols −, =, + × and ÷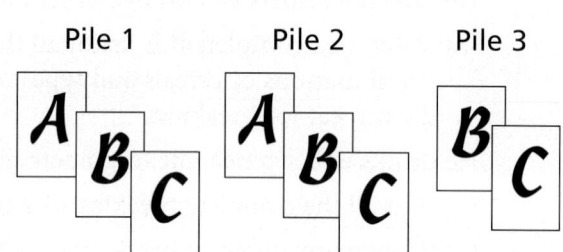
They are shuffled and placed face down.

  A card is chosen at random, and the symbol noted.
The card is replaced and they are shuffled again.
A second card is chosen at random and recorded.

   (a) List all the possible outcomes of the 2 cards.

   (b) What is the probability that both cards are the same?

   (c) What is the probability that one card is + and the other card is - ?

3 8 cards are put in 3 piles as
shown in the picture.

  The cards are placed in their piles
face down and shuffled.

  A card is chosen at random from each pile.

   (a) List all the possible outcomes

  What is the probability that:

   (b) the cards show all the same letter

   (b) the cards have 2 letters the same

   (c) the letters are all different

   (d) there is exactly 1 A

   (e) there are exactly 2 A's

Pile 1   Pile 2   Pile 3

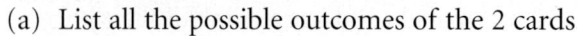
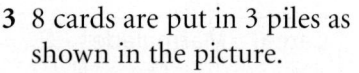

Section D

1 A dice numbered 1, 2, 3, 4, 5, 6 and a dice numbered 1, 1, 3, 3, 5, 6 are rolled and the numbers showing on the dice are added together.

(a) Copy and complete this table showing all the possible totals.

(b) What is the probability of getting a total of 2?

(c) What is the probability of scoring more than 6?

(d) Which is the most likely total?

(e) What is the probability of this total?

First dice

+	1	2	3	4	5	6
1						
1						
3		5				
3						
5						
6						

Second dice

2 Matthew spins a 5 sided spinner numbered 1, 2, 3, 4 and 5 and throws an ordinary dice.

He finds the **difference** in the numbers to get his score.

(a) Copy and complete the table to show all the possible scores.

(b) What is the probability of scoring 5?

(c) What is the probability that the score is greater than 3?

(d) What score is most likely and what is its probability?

(e) What is the probability of scoring 1 or 2?

Spinner

−	1	2	3	4	5
1				3	
2					
3		1			
4					
5					
6					

Dice

3 A 5 sided spinner numbered 1, 1, 2, 3 and 4 and another numbered 1, 2, 3, 4 and 4 are spun.

The numbers are multiplied together to get a score.

(a) Copy and complete this table

What is the probability that the score is

(b) less than 4?

(c) an even number?

(d) greater than 10?

First spinner

+	1	1	2	3	4
1					
2					
3					
4			8		
4					

Second spinner

Mixed practice 5

1 Write each of these lists in order, smallest first

   (a) $\frac{1}{2}$, 65%, 0.25, 20%, $\frac{3}{4}$ (b) 0.15, 10%, $\frac{1}{3}$, 20%, $\frac{1}{4}$

2 Work out

   (a) 10% of 60   (b) 20% of 60   (c) 20% of 80   (d) 25% of 60   (e) 10% of 500

3 Write in order, lowest first

   (a) 2°C, ⁻3°C, 1°C, ⁻6°C, 5°C   (b) 2.5°C, ⁻1.5°C, 0°C, 3.8°C, ⁻3.2°C

4 From the numbers in the loop, find two numbers that add to

   (a) 1   (b) 2   (c) ⁻2   (d) 0

                                                                          ⁻4 ⁻2 ⁻1 2 6

5 Work out   (a) ⁻3 × 3   (b) 3 × ⁻2   (c) 10 + 3x when x = ⁻4

6 On centimetre square paper, draw a grid with x and y going from 0 to 7.
   Then draw the parallelograms with these coordinates and find their areas.

   (a) (0, 0) (3, 0) (4, 2) (1, 2)   (b) (5, 1) (7, 3) (7, 7) (5, 5)

   (c) (0, 2) (4, 5) (4, 7) (0, 4)

7 What is the weight of each of these animals?

   (a)   (b)   (c)

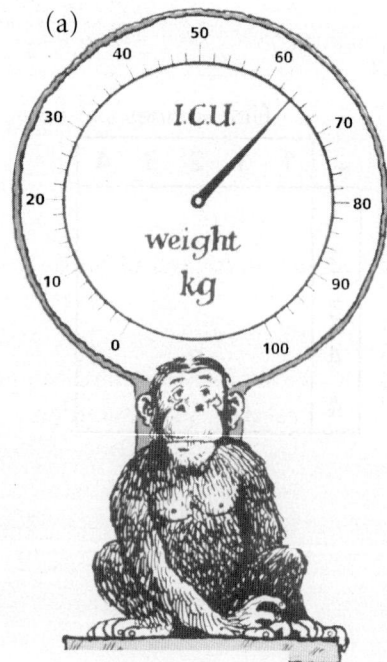

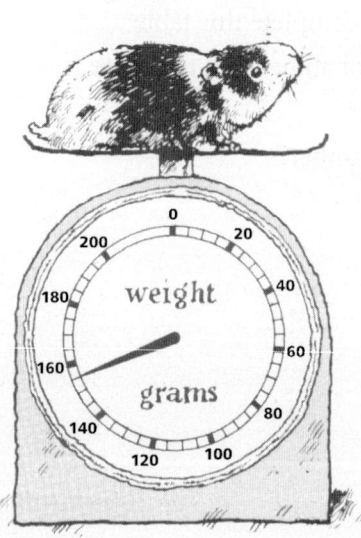

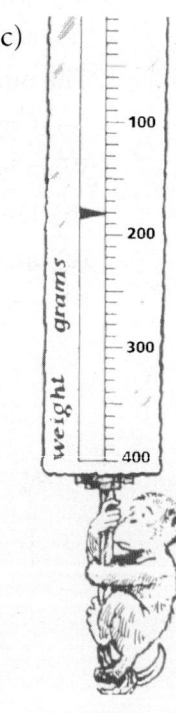

8 Write down and simplify an expression for the perimeter of each of these.

(a)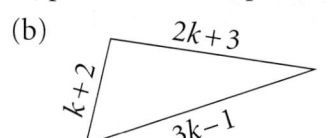

(b)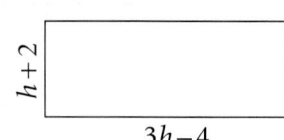

(c)

9 The scattergraph shows the length and weight of 8 eels caught in the River Severn.

(a) How long was the longest eel caught?

(b) How much did it weigh?

(c) Describe the type of correlation between the length and weight of these eels.

(d) Is the correlation strong or weak?

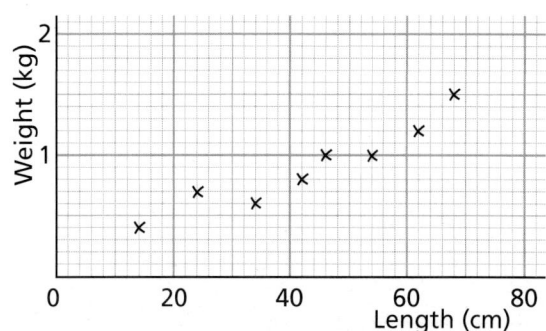

10 Work these out. Show your method clearly.

(a) 28×54 (b) 206×65 (c) $676 \div 13$ (d) $924 \div 44$

11 A packet of 24 biscuits costs 45p

(a) How much would 36 packets cost?   (b) How many biscuits are in 35 packets?

(c) How many packets can you buy for £9?

12 Jo has a normal dice and a coin.
One side of the coin has a 1 written on it. The other side has a 2 on it.

Jo rolls the dice and flips the coin.
She adds the two numbers together to get the score.

(a) Copy and complete this table to show all the possible scores.

(b) What is the probability that the score is 5?

(c) What is the probability that the score is 4 or less?

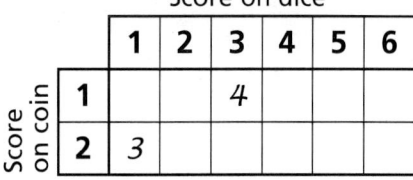

13 (a) You come out of St Mary's hospital and turn left into Praed Street. Which compass direction is that?

(b) From Talbot Square you go north east into Sussex Gardens. You take the second left and first right. Where are you?

(c) Give directions to get from Paddington Tube station (marked ⊖) to Radnor Place.

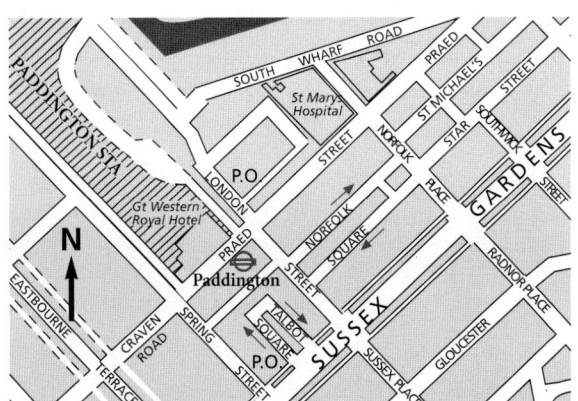